ERRATUM

Amor Sacro e Amor Profano
oil on canvas (118 x 279 cm)
Tiziano Vecellio (1514 ca.)
Galleria Borghese, Rome.

The painting was commissioned for the wedding of Laura Bagarotto to Nicolo Aurelio (1514), and it shows the bride assisted by Venus with Amor in the middle. Contrary to the impression given by the title, which would relate the dressed figure to Sacred Love and Worldly Love to the nude, the painter wanted to illustrate the continuity (duality, we might say) of love of individual terrestrial beauty (the bride holding the cup of joy) to love of the beauty of the cosmic order (represented by the goddess and the eternal flame she holds). The different landscapes behind the figures reinforce the meaning.

PROCEEDINGS OF THE FIRST WORKSHOP ON
QUARK–HADRON DUALITY
AND THE TRANSITION TO pQCD

PROCEEDINGS OF THE FIRST WORKSHOP ON
QUARK–HADRON DUALITY
AND THE TRANSITION TO pQCD

Frascati, Italy 6 – 8 June 2005

Alessandra Fantoni *(INFN-Frascati, Italy)*
Simonetta Liuti *(University of Virginia, USA)*
Oscar A. Rondón *(University of Virginia, USA)*

editors

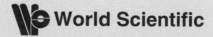

World Scientific

NEW JERSEY · LONDON · SINGAPORE · BEIJING · SHANGHAI · HONG KONG · TAIPEI · CHENNAI

Published by

World Scientific Publishing Co. Pte. Ltd.

5 Toh Tuck Link, Singapore 596224

USA office: 27 Warren Street, Suite 401-402, Hackensack, NJ 07601

UK office: 57 Shelton Street, Covent Garden, London WC2H 9HE

British Library Cataloguing-in-Publication Data
A catalogue record for this book is available from the British Library.

QUARK–HADRON DUALITY AND THE TRANSITION TO pQCD
Proceedings of the First Workshop

ISBN 981-256-684-8

Printed in Singapore by World Scientific Printers (S) Pte Ltd

CONTENTS

PREFACE

This Volume contains the invited talks and contributed papers presented at the "First Workshop on Quark-Hadron Duality and the Transition to pQCD", that took place in the Laboratori Nazionali di Frascati (Rome), Italy, in June 2005.

The aim of the Workshop was to discuss recent results, and to foster current and future research on the phenomenon of quark-hadron duality.

Understanding the structure of hadrons, and their long distance interactions in terms of quark and gluon degrees of freedom is probably the most challenging question for Quantum Chromodynamics (QCD). The main challenge resides in the fact that hadrons, as bound systems of quarks and gluons, are described within the strong coupling regime of QCD which, in turn, is responsible for the chiral symmetry breaking and confinement phenomena. On the other side, at short distances or in the weak coupling regime, the quark and gluon structure of hadrons is revealed by using high energy probes, such as in deep inelastic scattering experiments. The trasmogrification into hadrons happens at a much later time scale to be able to influence the cross section. As the energy of the probe is decreased towards values closer to typical hadronic scales, the effects of confinement are expected to dominate the cross section, which displays a resonance structure generally interpreted in terms of collective degrees of freedom, *i.e.* mesons and baryons. Clearly, a future solution to the theory of strong interactions will find the descriptions in terms of hadronic and partonic degrees of freedom to be equivalent. In many circumstances it is has been, in fact, already possible to observe similarities between properly averaged hadronic cross sections, and the partonic ones. This is the phenomenon of quark-hadron duality that reflects the relationship between confinement and asymptotic freedom, being intimately related to the nature of the transition from non-perturbative (nPQCD) at low energy, to perturbative QCD (PQCD) at high energy.

One of the most illustrative examples is given by electron-nucleon scattering, where the low-energy cross sections, averaged over properly defined energy intervals, are found to exhibit the scaling behavior expected from perturbative QCD. Depending on the choice of energy intervals, or averaging procedures in the resonance region, different definitions of duality have been given: one can in fact refer to *global* duality if the average, or the integral of the structure functions, is taken over the whole resonance region. If, however, the averaging is performed over smaller ranges, extending *e.g.* over single resonances, one can refer to *local* duality.

Although duality between the quark and hadronic descriptions is expected to be in principle a universal relationship, how it reveals itself specifically in different physical processes, and under different kinematical conditions is also a crucial point for understanding the hadronic structure of QCD. The phenomenon of duality was observed in a number of processes, from deep inelastic scattering, to e^+e^- annihilation into hadrons, hadron-hadron collisions, and semi-leptonic decays of heavy quarks. Both recent theoretical progress, and higher precision measurements covering a wide range of reactions, are now making it possible to investigate the role of duality in QCD as a subject in itself.

The Workshop succeeded in bringing together for the first time researchers from different areas of hadronic physics, presenting and discussing the latest results on different manifestations of quark-hadron duality and addressing as one group this exciting topic.

The initial Sessions included reviews on the status of theoretical and experimental studies. Recent progress on structure functions was presented in detail also in additional contributed talks: the latest results were shown and discussed for both unpolarised and polarised scattering in different kinematical regimes, including results on the Gerasimov-Drell-Hearn sum rule which is crucial in this context. Theoretical talks addressed different aspects of non-perturbative approaches: duality and quark models, lattice QCD calculations of the strong coupling constant, chiral extrapolations of lattice QCD results. Aspects of the transition from the non-perturbative to the perturbative regimes of QCD were discussed in one dedicated session. In particular applications of the AdS/CFT hypothesis to hadron spectra and to several aspects of structure functions such as dimensional counting rules were presented. Using QCD sum rules, the interpretation of quark-hadron duality in terms of field theoretical aspects of QCD was introduced. The latter is related to the possibility of extracting accurately higher twist effects, which play an important role in the nPQCD to PQCD transition, from both the resonance and the deep inelastic regions and in different reactions, a topic addressed specifically in this workshop. Finally, the hypothesis and nature of the existence of QCD scales different from the one related to the typical hadronic size was presented. Duality was analysed also in a number of different reactions: photoproduction, using nuclei, including also a discussion on hadronization and the quark-gluon plasma, and for neutrino scattering. Finally, the Workshop ended with a session on future perspectives on both theoretical and experimental fronts. The topic of Generalized Parton Distributions (GPDs) was discussed, identifying a new formalism directed at a unified description of hard exclusive and inclusive processes. The main issues to be addressed in the next ten years at the upcoming and proposed new facilities were presented including the 12 GeV upgrade of Jefferson Laboratory, the approved experiment, PANDA, and the proposed one, PAX, both at GSI.

It is our pleasure to acknowledge the support of the institutional sponsors that made this Workshop possible: the Istituto Nazionale di Fisica Nucleare, the Thomas Jefferson National Accelerator Facility, and Hampton University.

We also gratefully acknowledge the advice from the members of the International Advisory Committe. Our special thanks go to all participants, speakers and conveners who contributed to the success of this lively and interesting meeting.

The Editors
Alessandra Fantoni
Simonetta Liuti
Oscar A. Rondon

ERRATUM

WORKSHOP ORGANIZATION

International Advisory Committee

S. Bertolucci (LNF - Italy)
I. Bigi (Notre Dame University - France)
S. Brodsky (SLAC - USA)
V. Burkert (JLAB - USA)
F. Close (Oxford University - United Kingdom)
K. de Jager (JLAB - USA)
E. De Sanctis (LNF - Italy)
Y.L. Dokshitzer (Paris University - France; LPTH, St.Petersburg - Russia)
R. Ent (JLAB - USA)
P. Hoyer (Helsinki University - Finland)
X. Ji (Maryland University - USA)
P. Kroll (Wuppertal University - Germany)
A. Lung (JLAB - USA)
Z. Meziani (Temple University - USA)
L. Pancheri (LNF - Italy)
J.C. Peng (Illinois University - USA)
A.V. Radyushkin (JLAB, Old Dominion University - USA; JINR - Russia)
K. Rith (Erlangen University - Germany)
M.J. Savage (Washington University, Seattle University - USA)
J. Soffer (Marseille University - France)
M. Taiuti (University of Genova - Italy)
A.W. Thomas (JLAB, USA)
G. Van der Steenhoven (NIKHEF - Netherlands)

International Organizing Committee

N. Bianchi (LNF)
J.P. Chen (JLAB)
P. Di Nezza (LNF)

A. Fantoni (LNF)
C. Keppel (Hampton University)
S. Liuti (University of Virginia)
V. Muccifora (LNF)
F. Ronchetti (LNF)
O. Rondon (University of Virginia)

Workshop Chairpersons

N. Bianchi (LNF)
J.P. Chen (JLAB)
A. Fantoni (LNF)
S. Liuti (University of Virginia)

Secretary

D. Pierluigi (LNF)

Web Manager

F. Ronchetti (LNF)

Institutional Sponsors

Istituto Nazionale di Fisica Nucleare
Jefferson Laboratory
Hampton University

Introduction and Review:
Experimental and Theoretical Status

INTRODUCTORY REMARKS ON DUALITY IN LEPTON-HADRON SCATTERING

C. E. CARLSON

Particle Theory Group, Physics Department
College of William and Mary, Williamsburg, VA 23187-8795, USA

We consider some aspects of duality in lepton-hadron inelastic scattering, including how duality works when perturbative QCD is applicable, how to explain the peculiar behavior of the $\Delta(1232)$, and the finding that duality also appears in quantum mechanical models with confinement. We include an example of an application of duality-pertinent data to atomic physics, specifically to proton structure corrections to hydrogen hyperfine splitting.

1. Introduction

Duality in electron-hadron scattering is a statement that the resonance bumps seen in inelastic data at low Q^2 average out—if one uses the correct variables—to the scaling curve that is seen at higher Q^2. Duality has been known since the work of Bloom and Gilman[1] in 1970. We show here a 1991 figure from Stoler[2], where one can see at several values of Q^2 the resonance bumps in data plotted vs. (essentially) $1/x$ where x is the scaling variable $x = Q^2/2m_p\nu$. The solid curve is the scaling curve plotted as a function of the scaling variable, but measured at higher Q^2. One can see that the averaging is working, at least approximately, and that it continues to work as Q^2 changes and the resonance bumps appear at different values of x and slide down the scaling curve. Modern times have provided us with much new data, and also an excellent full length review article[3].

2. Duality when PQCD is valid

I would like to begin the more detailed discussion by showing how the resonance to continuum ratio maintains its constancy at all (large enough) Q^2. There was, to be sure, a "proof", or at least a "demythification," of duality offered by DeRújula, Georgi, and Politzer[4] in 1977. However, I would prefer a more explicit understanding[5].

4

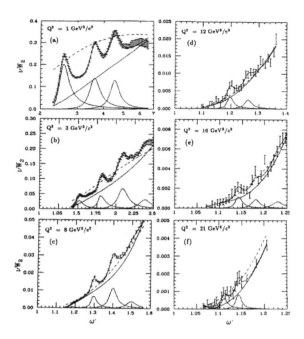

Figure 1. A plot from[2] of data from inelastic electron-proton scattering. The six values of Q^2 shown above are labeled on the graphs. The solid curve is the scaling curve plotted on each graph as a function of ω', but measured at higher Q^2. ($1/\omega' \approx x = Q^2/2m_p\nu$.)

Let me talk directly in terms of resonance production, and phrase the discussion using the helicity amplitudes or helicity form factors $G_\pm$ and G_0. Treating the resonances as stable particles, the helicity matrix elements are,

$$G_m = \left\langle R, \lambda' = m - \frac{1}{2} \middle| \epsilon_m \cdot J \middle| N, \lambda = \frac{1}{2} \right\rangle \middle/ (2m_N) \qquad (1)$$

(For the elastic case, we would have (with τ defined farther below)

$$G_0 = G_E, \qquad G_+ = \sqrt{2\tau}G_M, \quad \text{and} \quad G_- = 0. \quad) \qquad (2)$$

The differential cross section for resonance electroproduction, using a Breit-Wigner form for the propagator to account for the fact that the physical resonance is unstable, is given by

$$\frac{d\sigma_R}{d\Omega_{lab}\,dx} = \frac{\sigma_{NS}}{1+\tau}\frac{\tau}{\pi}\frac{4m_N^2 m_R \Gamma_R}{\left(W^2 - m_R^2\right)^2 + m_R^2 \Gamma_R^2}\left(G_0^2 + \frac{1}{2\epsilon}\left(G_+^2 + G_-^2\right)\right) \qquad (3)$$

with $1/\epsilon = 1 + 2(1+\tau)\tan^2(\theta/2)$, $\tau = \nu^2/Q^2 = (Q^2/4m_p^2)(1/x^2)$, and σ_{NS} is the no-structure differential cross section which is well known.

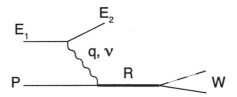

Figure 2.   Electroproduction of resonances.

At the resonance peak $(W = m_R)$ this is

$$\frac{d\sigma_R}{d\Omega_{lab}\,dx} = \frac{\tau\,\sigma_{NS}}{1+\tau}\,\frac{4m_N^2}{\pi m_R \Gamma_R}\left(G_0^2 + \frac{1}{2\epsilon}\left(G_+^2 + G_-^2\right)\right) \qquad (4)$$

We compare to the general form of the deep inelastic scattering cross section, in the form

$$\frac{d\sigma_{DIS}}{d\Omega_{lab}\,dx} = \frac{\tau\,\sigma_{NS}}{1+\tau}\,\frac{1}{x}\left(F_L + \frac{1}{\epsilon}F_T\right), \qquad (5)$$

where the transverse and longitudinal structure functions are related to the more common F_1 and F_2 structure functions by

$$F_T(x, Q^2) = 2xF_1(x, Q^2)$$
$$F_L(x, Q^2) = \left(1 + \frac{1}{\tau}\right)F_2(x, Q^2) - 2xF_1(x, Q^2) \qquad (6)$$

Hence for $x \to 1$

$$F_T \propto G_+^2 + G_-^2,$$
$$F_L \propto G_0^2. \qquad (7)$$

The left-hand-side depends on x only, in the scaling limit, while the right-hand-side depends on Q^2 only. They are correlated because we fix $W = m_R$,

$$W^2 = (P+q)^2 = m_N^2 + 2m_N\nu - Q^2 \quad \text{or} \quad (1-x) = \frac{m_R^2 - m_N^2}{Q^2} \qquad (8)$$

the latter for $x \to 1$.

The counting rules, which come from perturbative QCD and the knowledge that baryons are made from 3 quarks, tell us that

$$G_+^2 \propto Q^{-6} \qquad\qquad G_0^2 \propto Q^{-8} \qquad\qquad G_-^2 \propto Q^{-10} \qquad (9)$$

This in turn says that as the resonances slide down the curves describing $F_{T,L}$, they slide along curves

$$F_T \propto (1-x)^3 \qquad\qquad F_L \propto (1-x)^4 \qquad\qquad (10)$$

But this is what we know that F_T and F_L do anyway, from their own counting rules (and verified from data to some decent approximation). A similar analysis works[6,7] for the polarized structure function g_1.

2.1. *The behavior of the Delta(1232)*

The $\Delta(1232)$, unlike resonances in the 1535 or 1688 MeV regions, becomes progressively harder to find as Q^2 increases; look at Fig. 1. One would like to know why the Δ is different.

The reason the Δ disappears is that the asymptotic size of the leading helicity form factor is anomalously small. This is not just an inference from the observation, but also a result of a pQCD calculation, similar to the better known pQCD calculation of the high Q^2 nucleon elastic dirac form factor F_1. Hence what we mainly see in $N \to \Delta$ are asymptotically subleading amplitudes. It is a lousy circumstance for pQCD that the first resonance is an exceptional case, yet we can repeat that it is a circumstance substantiated by calculation.

However—and I consider this a big "however"—duality is still maintained is the $\Delta(1232)$ region[8]. Duality means that the average over the resonance region matches the the average using the scaling curve. It does— even for $\Delta(1232)$. What happens is that as the resonance peak falls, the background rises, and average/continuum $\approx$ constant. One concludes that, somehow, the background knows about the Δ, and co-concludes that one should not use just simple π-nucleon Born terms to model the background.

3. Modeling scaling from quantum mechanics and bound state models

Several sets of workers, including[9,10,11], have in recent times developed quantum mechanical models that show features connecting bound state and continuum behavior. These models may be useful in understanding duality in the physical world. The ingredients of the models are bound states in quantum mechanics, pointlike particles, and a confining potential.

Consider a bound state that starts in the ground state, and then gets hit and transits to an excited state, with some transition form factor. One can use an exactly solvable relativistic harmonic oscillator to get definite results for the transition form factor, such as

$$F_{0 \to N} = \frac{i^N}{\sqrt{N!}} \left(\frac{|\vec{q}|}{\beta \sqrt{2}} \right)^N e^{-\vec{q}^2/4\beta^2} \tag{11}$$

The transition form factor is small at low and high $\vec{q}^2$, and peaks at some $\vec{q}^2$ that happens to be at the same x_{Bj}, or x_{Bj} equivalent, for each transition form factor regardless of the final state. The practitioners of at least one model[10] prefer to define the scaling variable as

$$u \equiv \frac{m}{M} x_{Bj} \tag{12}$$

where in this model m is the mass of the light struck quark and M is the mass of the heavy quark it is bound to.

The response function is

$$\frac{d\sigma}{dE_f \, d\Omega_f} \propto \mathcal{S} \sim \sum |F_{0 \to N}|^2 \, \delta(E_N - E_0 - \nu) \tag{13}$$

which is a collection of delta-functions, with the delta-functions having different multiplicative factors and different locations in the scaling variable u for different values of momentum transfer Q^2. They can be given some artificial width for visual purposes, and then one gets a set of curves for different momentum transfers that looks like Fig. 3.

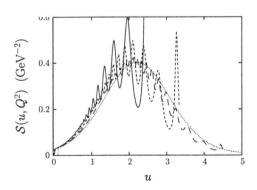

Figure 3. The response function for different values of Q^2 in one quantum mechanical model[10] showing duality between resonance excitations and a scaling curve. The values of Q^2 are $Q^2 = 0.5$ (solid), 1 (short-dashed), 2 (long-dashed), and 5 GeV^2 (dotted). The figure is from .

The limiting curve, amazingly enough, is the same as one would get from the initial state wave function, treating the final "quarks" as free.

4. Proton structure and atomic hyperfine splittings

Here is a piece of "applied physics," in the sense of using for another field of physics some of the quantities that have been measured and used in duality studies. The other field is atomic physics, and the quantities in question are elastic form factors and $g_{1,2}$ on the hadronic side and proton structure corrections to hyperfine splittings on the atomic side.

Atomic hyperfine splittings (hfs) are very accurately measured. For hydrogen and muonium one has[12],

$$E_{\text{hfs}}(ep) = 1\ 420.405\ 751\ 766\ 7(9)\ \text{MHz},$$
$$E_{\text{hfs}}(e\mu) = 4\ 463.302\ 78(5)\ \text{MHz}. \tag{14}$$

which are 13 figures and 8 figures accurate, respectively.

The leading order calculation of the splitting is due to Fermi,

$$E_F = \frac{8}{3\pi} \alpha^3 \mu_B \mu_N \frac{m_e^3 m_N^3}{(m_N + m_e)^3} \tag{15}$$

where "N" stands for either p or μ^+ and the μ_i's are magnetic moments. There are corrections, as

$$E_{\text{hfs}}(ep) = E_F(ep) \times (1 + \Delta_{\text{QED}} + \Delta_Z + \Delta_{\text{pol}} + \Delta_R) \tag{16}$$

where Δ_{QED} is the same for hydrogen and muonium, Δ_R is the recoil and radiative recoil correction (also really QED), $\Delta_Z + \Delta_{\text{pol}}$ together are proton structure corrections, and Δ_Z is a purely elastic part of the correction, worked out by Zemach many years ago[13],

$$\Delta_Z = -2\alpha m_e \langle r \rangle_Z \times \left(1 + \delta^{\text{radiative}}\right). \tag{17}$$

Here $\langle r \rangle_Z$ is the "Zemach radius,"

$$\langle r \rangle_Z = -\frac{4}{\pi} \int_0^\infty \frac{dQ}{Q^2} \left[G_E(Q^2) \frac{G_M(Q^2)}{\mu_p} - 1 \right], \tag{18}$$

with $G_M(0) = \mu_p$, and $\delta^{\text{radiative}}$ is a known $\mathcal{O}(\alpha)$ correction ($\approx 1.53\%$).

The polarizability corrections come mainly from inelastic intermediate states (e.g.,[14]),

$$\Delta_{\text{pol}} = \frac{\alpha m_e}{\pi g_p m_p} (\Delta_1 + \Delta_2),$$

$$\Delta_1 = \frac{9}{4} \int_0^\infty \frac{dQ^2}{Q^2} \left\{ F_2^2(Q^2) + 4m_p \int_{\nu_{\text{th}}}^\infty \frac{d\nu}{\nu^2} \beta \left(\frac{\nu^2}{Q^2} \right) g_1(\nu, Q^2) \right\},$$

$$\Delta_2 = -12 m_p \int_0^\infty \frac{dQ^2}{Q^2} \int_{\nu_{\text{th}}}^\infty \frac{d\nu}{\nu^2} \beta_2 \left(\frac{\nu^2}{Q^2} \right) g_2(\nu, Q^2), \tag{19}$$

where F_2 is the Pauli form factor, g_1 and g_2 are spin-dependent structure functions, and

$$\beta(\tau) = \frac{4}{9}\left(-3\tau + 2\tau^2 + 2(2-\tau)\sqrt{\tau(\tau+1)}\,\right),$$
$$\beta_2(\tau) = 1 + 2\tau - 2\sqrt{\tau(\tau+1)}\ .$$

Faustov and Martynenko[15] in 2002 evaluated $\Delta_{\text{pol}} = 1.4 \pm 0.6$ ppm.

(As an aside, note that Δ_1 is finite by dint of the GDH sum rule[16,17].)

One can use these calculated corrections is several ways.

• Use the proton hfs alone and the calculated Δ_{QED} and infer $\Delta_Z + \Delta_{\text{pol}}$.

• Or, use the difference of proton and muonium hfs to eliminate the (big) calculated QED correction[18], and infer $\Delta_Z + \Delta_{\text{pol}}$.

• Either way, one can use the presently calculated Δ_{pol} to infer Δ_Z or $\langle r \rangle_Z$, obtaining

$$\langle r \rangle_Z = 1.043(16)\ \text{fm}\,, \tag{20}$$

which then becomes a constraint on any form factor parameterization.

• Or one can take the "best" form factor knowledge to calculate $\langle r \rangle_Z = 1.086(12)$ fm, and infer[19] $\Delta_{\text{pol}} = 3.05(49)$.

With either of the last two choices, Δ_{pol} is a weak point. Its evaluation is sensitive to $g_{1,2}$ at low Q^2. Newer data with Q^2 down to 0.05 GeV^2 (about a factor of 3 lower than the data[20] published in 2003) proves important. A re-evaluation of Δ_{pol} is underway[21]. Presently, we can report that the central value seems somewhat smaller—not larger—than the old result[15], and the uncertainty limit can be about half of what Ref.[15] found.

5. Final remarks

Allow me to briefly repeat things that have been said earlier. We have seen that both data and theory gives us reason to think duality works when pQCD is applicable. Further, duality seems also to appear in quantum mechanical models with confinement. Nonetheless, we still would like to have a more general understanding and how and how well duality works, particularly since we want to use duality in applications like studying the structure functions for $x_{Bj} \to 1$ by using data in the resonance region.

We have looked at one example of a useful application of duality-pertinent data in a related area, namely in calculating proton structure corrections to hydrogen hyperfine splitting.

Finally, although something we did not discuss, we can expect duality in other hadronic physics processes, for example in semiexclusive processes.

Acknowledgments

Thanks to the organizers for an excellent conference, also to the National Science Foundation for support under Grant No. PHY-0245056.

References

1. E. D. Bloom and F. J. Gilman, Phys. Rev. D **4**, 2901 (1971); Phys. Rev. Lett. **25**, 1140 (1970).
2. P. Stoler, Phys. Rev. D **44**, 73 (1991). Phys. Rev. Lett. **66**, 1003 (1991).
3. W. Melnitchouk, R. Ent and C. Keppel, Phys. Rept. **406**, 127 (2005) [arXiv:hep-ph/0501217].
4. A. De Rujula, H. Georgi and H. D. Politzer, Annals Phys. **103**, 315 (1977).
5. C. E. Carlson and N. C. Mukhopadhyay, Phys. Rev. D **41**, 2343 (1990).
6. C. E. Carlson and N. C. Mukhopadhyay, Phys. Rev. D **58**, 094029 (1998) [arXiv:hep-ph/9801205].
7. X. D. Ji and J. Osborne, J. Phys. G **27**, 127 (2001) [arXiv:hep-ph/9905410].
8. C. E. Carlson and N. C. Mukhopadhyay, Phys. Rev. D **47**, 1737 (1993). Phys. Rev. Lett. **81**, 2646 (1998) [arXiv:hep-ph/9804356].
9. M. W. Paris and V. R. Pandharipande, Phys. Lett. B **514**, 361 (2001) [arXiv:nucl-th/0105076].
10. N. Isgur, S. Jeschonnek, W. Melnitchouk and J. W. Van Orden, Phys. Rev. D **64**, 054005 (2001) [arXiv:hep-ph/0104022].
11. S. A. Gurvitz and A. S. Rinat, Phys. Rev. C **65**, 024310 (2002) [arXiv:nucl-th/0106032].
12. S. G. Karshenboim, Can. J. Phys. **77**, 241 (1999).
13. A. C. Zemach, Phys. Rev. **104**, 1771 (1956).
14. S. D. Drell and J. D. Sullivan, Phys. Rev. **154**, 1477 (1967).
15. R. N. Faustov and A. P. Martynenko, Eur. Phys. J. C **24**, 281 (2002); R. N. Faustov and A. P. Martynenko, Phys. Atom. Nucl. **65**, 265 (2002) [Yad. Fiz. **65**, 291 (2002)].
16. S. B. Gerasimov, Sov. J. Nucl. Phys. **2**, 430 (1966) [Yad. Fiz. **2**, 598 (1966)].
17. S. D. Drell and A. C. Hearn, Phys. Rev. Lett. **16**, 908 (1966).
18. S. J. Brodsky, C. E. Carlson, J. R. Hiller and D. S. Hwang, Phys. Rev. Lett. **94**, 022001 (2005); Phys. Rev. Lett. **94**, 169902 (E) (2005) [arXiv:hep-ph/0408131]. See also [19].
19. J. L. Friar and I. Sick, Phys. Rev. Lett. **95**, 049101 (2005) [arXiv:nucl-th/0503020] and S. J. Brodsky, C. E. Carlson, J. R. Hiller and D. S. Hwang, Phys. Rev. Lett. **95**, 049102 (2005).
20. R. Fatemi *et al.* [CLAS Collaboration], Phys. Rev. Lett. **91**, 222002 (2003) [arXiv:nucl-ex/0306019].
21. K. Griffioen, V. Nazaryan, and C. E. Carlson, forthcoming.

DUALITY IN THE POLARIZED STRUCTURE FUNCTIONS

H.P. BLOK

Vrije Universiteit and NIKHEF
Amsterdam
E-mail: henkb@nikhef.nl

The special features of the polarized structure functions g_1 and g_2 in relation to duality are briefly discussed, and the present knowledge about g_1 and g_2 in the resonance region is reviewed.

1. Introduction

For an introduction to and overview of duality, see [refciteCarls,PhRep. The signature of duality is that the average over (some) resonances of a cross section or structure function, measured in the resonance region (hence at relatively low Q^2), is equal to the 'scaling' curve, i.e., the value in the deep-inelastic (DIS) region at the same value of Bjorken-x (thus at relatively high Q^2). The interest in duality from the theoretical side stems from the fact that it can tell something about the QCD structure of the nucleon and its excited states and higher twist contributions, while if duality is fulfilled, it allows one to measure structure functions in the important valence region to large(r) values of x.

Obvious questions are: what range does one have to average over (how 'local' or 'global' is duality), and what is the definition of the scaling curve? Since the latter is at higher Q^2, how does one take into account the difference in Q^2 (evolution, target mass corrections, other effects)? One should look also to differences between protons and neutrons, longitudinal and transverse structure functions, spin dependence, etc. Refs. [4, 5 give an illuminating insight how within an SU(6) quark model duality can come about, and what the crucial ingredients are.

2. Spin dependence

In order to determine the spin dependent structure functions g_1 and g_2 one measures the cross sections for inclusive (e,e') scattering with a longi-

11

tudinally polarized electron and a longitudinally or transversely polarised target, and calculates the lepton asymmetries, i.e., with respect to the beam axis:

$$A_{\parallel} = \frac{\sigma^{\leftarrow \rightarrow} - \sigma^{\rightarrow \rightarrow}}{\sigma^{\leftarrow \rightarrow} + \sigma^{\rightarrow \rightarrow}}$$

$$A_{\perp} = \frac{\sigma^{\leftarrow \perp} - \sigma^{\rightarrow \perp}}{\sigma^{\leftarrow \perp} + \sigma^{\rightarrow \perp}}. \qquad (1)$$

From these one can deduce[5] the photon asymmetries (i.e., with respect to the virtual photon axis):

$$A_1 = \frac{A_{\parallel}}{D(1 + \eta\zeta)} - \frac{\eta A_{\perp}}{d(1 + \eta\zeta)} = \frac{g_1(x, Q^2) - \gamma^2 g_2(x, Q^2)}{F_1(x, Q^2)}$$

$$A_2 = \frac{\zeta A_{\parallel}}{D(1 + \eta\zeta)} + \frac{A_{\perp}}{d(1 + \eta\zeta)} = \frac{\gamma[g_1(x, Q^2) + g_2(x, Q^2)]}{F_1(x, Q^2)}. \qquad (2)$$

Here D and d are kinematic factors of order unity, and η and ζ are of order $\gamma = Q/\nu$, which for measurements in the DIS region usually is rather small, but is non-negligible in the resonance region.

As is well known A_1 and g_1 for the proton are positive for large values of Q^2. However, at small values of Q^2 g_1 is found to be negative in the resonance region around the $\Delta(1232)$ peak. This means that duality (at least locally) must fail below a certain Q^2. This is related to the behaviour of the generalised GDH integral[6,?]:

$$I_{GDH}(Q^2) = \int_{\nu_0}^{\infty} [\sigma_{1/2}(\nu, Q^2) - \sigma_{3/2}(\nu, Q^2)] \frac{d\nu}{\nu}$$

$$= \frac{8\pi^2 \alpha}{M} \int_0^{x_0} \frac{g_1(x, Q^2) - \gamma^2 g_2(x, Q^2)}{K} \frac{dx}{x}, \qquad (3)$$

where ν_0 corresponds to the threshold for pion production, $\sigma_{1/2}$ and $\sigma_{3/2}$ are the helicity dependent virtual-photon absorption cross sections, and K is the equivalent photon energy. At large Q^2 $I_{GDH}(Q^2)$ is positive and related to the Ellis-Jaffe sum rule:

$$\Gamma_1(Q^2) = \int_0^1 g_1(x, Q^2) dx = \frac{Q^2}{16\pi^2 \alpha} I_{GDH}(Q^2), \qquad (4)$$

whereas for $Q^2 \to 0$

$$I_{GDH}(Q^2) = -\frac{2\pi^2 \alpha}{M^2} \kappa^2 \qquad (5)$$

(with κ the anomalous magnetic moment), which is negative. Thus the integral is strongly Q^2 dependent (at least for the proton, for the neutron it is always negative). Including the elastic peak overcompensates for this[2]. The behaviour for the proton is shown in Fig. 1. It is well described by a model[9] that explicitly takes the $\Delta(1232)$ into account. A model[10] using a (simple) parametrization of $g_1 + g_2$ is going to be improved[11]. So it seems that duality in spin-dependent structure functions is more sensitive, which is supported by arguments given in [4].

For g_2 two relations are of importance: the Burkhardt-Cottingham sum rule[12], which is supposed to be generally valid

$$\int_0^1 g_2(x, Q^2)dx = 0, \qquad (6)$$

and the twist-2 Wandzura-Wilczek relation[13]

$$g_2^{WW}(x, Q^2) = -g_1(x, Q^2) + \int_x^1 \frac{g_1(y, Q^2)}{y} dy, \qquad (7)$$

which seems to be a good approximation.

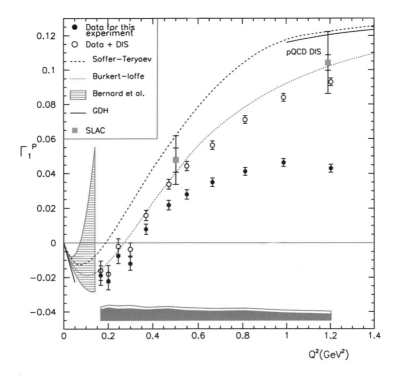

Figure 1. Q^2 dependence of $\Gamma_1(Q^2)$ for the proton (from [8]).

14

3. Existing data

The first measurements in the resonance region of the spin dependent structure functions were performed for both proton and deuteron at SLAC[5]. By using beam energies of 9.7, 16.2 and 29.1 GeV and polarized NH_3 and ND_3 targets both $A_\parallel$ and $A_\perp$ were measured at values of Q^2 of 0.5 and 1.2 $(GeV/c)^2$ for $W^2 < 5$ GeV^2, and for typical values of 3 $(GeV/c)^2$ in the DIS region. Some data are shown in Fig. 2. Especially for the proton at

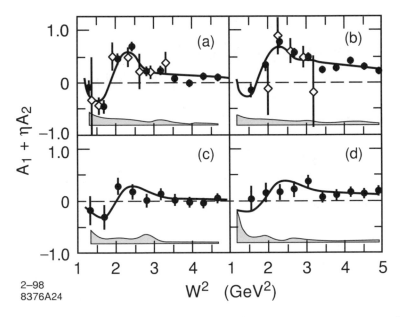

Figure 2. $A_\parallel/D$ data for the proton (top) and the deuteron (bottom) for $Q^2 = 0.5$ (left) and 1.2 $(GeV/c)^2$ (right). The curve is a parametrization of all existing data.

low Q^2 the asymmetry has a clear structure, with a negative peak in the $\Delta(1232)$-region and a positive one around $W^2 = 2.4$ GeV^2 where the S_{11} and D_{13} resonances are important. The asymmetry reaches values close to -0.5 and 1.0 in these two regions, consistent with[14,6] the expected dominant M_1 (M_{1+} pion multipole) and E_1 (E_{0+} pion multipole) transitions, resp. (it is interesting to note that the M_{1-} multipole, which plays a role in the background underneath the Δ and at higher values of W^2 has an asymmetry of +1.0). The data for g_2 were found to be consistent with Eq. 7.

The resonance region on the proton was also studied at Hermes. With 27.6 GeV polarized positrons on an open gas cell target fed by an Atomic Beam Source, $A_{\parallel}$ was measured[15]. By using the SLAC estimate for A_2, values for A_1 were determined in the region $x = 0.35 - 0.8$ at average values of Q^2 of $1.6 - 2.9$ (GeV/c)^2. Within the error bars the data points lie nicely on the curve $A_1 = x^{0.7}$, which is a fit over the world DIS data at high x ($x > 0.3$). For a more detailed study the integral

$$\Gamma_1^{res}(Q^2) = \int_{x_{min}}^{x_{max}} g_1^{res}(x, Q^2) dx \qquad (8)$$

was determined and compared to the same integral in the DIS regime [16] (assuming for the latter that A_1 is independent of Q^2, so all Q^2-dependence comes from F_1). The ratio is shown in Fig. 3 for three values of Q^2, together with values from the SLAC experiment. Clearly the ratio is not constant if $Q^2 < 1.5$ (GeV/c)^2. A detailed discussion and a comparison with the upolarised case is reported in [17,18].

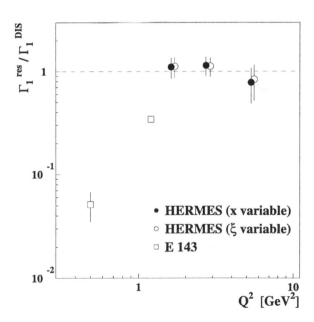

Figure 3. Ratio of Γ_1 for a certain x range in the resonance region to the one in the DIS region, as a function of Q^2.

The behaviour of g_1 in the resonance region has also been studied with CLAS at JLab[8,19]. Data for $A_{\parallel}$ for the proton and deuteron were measured in the region $Q^2 \approx 0.2 - 1.5$ (GeV/c)^2. With a model parametrization for A_2 values of g_1 were determined, as shown in Fig. 4 for the proton. There is a clear Q^2 dependent structure in the first, second and third resonance regions.

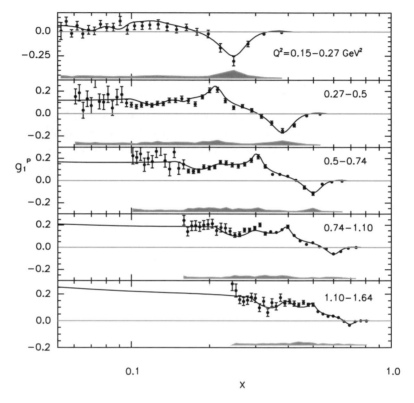

Figure 4. Data for g_1^p as a function of x for various Q^2.

Finally g_1 and g_2 for the neutron were measured in Hall A at JLab[20] for values of $Q^2 = 0.1 - 0.9$ (GeV/c)^2 by using a polarized ^{3}He target. Again a distinct structure in the Δ region was observed. Furthermore it was found that $g_2 \approx -g_1$ (see further [21]).

4. Conclusions and outlook

It seems that when $Q^2 > 1.5$ (GeV/c)^2 global duality for g_1 is fair (and better for the neutron than for the proton). Local duality is strongly violated in the Δ and $S_{11} + D_{13}$ regions, but is better obeyed beyond $W = 1.6$

GeV. The former has consequences for using duality to determine g_1 (and g_2) at high x, which may require the use of high (> 4 $(\text{GeV}/\text{c})^2$) values of Q^2, where the form factor of the Δ gets relatively small. It looks as if duality works better for the asymmetry A_1 than for the structure functions. Because of its relatively small value and the relatively large experimental uncertainties, very little is known about duality of g_2. It would be interesting to look for duality in the combination $g_1 + g_2$.

Given the evident role of the resonance form factors it would be good to look into duality for the resonance peaks and the background underneath them separately, and compare these to predictions from models for the resonance region. It seems that the background has some structure, not described by just Born terms.

All of this will benefit from more accurate data, which are becoming available from JLab, plus accompanying theoretical developments, see contributions in these proceedings.

References

1. C. Carlson, *These proceedings.*
2. W. Melnitchouk, R. Ent and C.E. Keppel, *Phys. Reports* **406**, 127 (2005).
3. F.E. Close and N. Isgur, *Phys. Lett.* **B509**, 81 (2001).
4. F.E. Close and W. Melnitchouk, *Phys. Rev.* **C68**, 035210 (2003).
5. K. Abe *et al.*, *Phys. Rev.* **D58**, 112003 (1998).
6. D. Drechsel and L. Tiator, *Ann. Rev. Nucl. Part. Sci.* **54**, 69 (2004).
7. A. Airapetian *et al.*, *Eur. Phys. Journal* **C26**, 527 (2003).
8. R. Fatemi *et al.*, *Phys. Rev. Lett.* **91**, 222002 (2003).
9. V.D. Burkert and B.L. Ioffe, *Phys. Lett.* **B296**, 223 (1992); V.D. Burkert and Zh. Li, *Phys. Rev.* **D47**, 46 (1993).
10. J. Soffer and O.V. Teryaev, *Phys. Rev.* **D51**, 25 (1995).
11. O.V. Teryaev, *These proceedings.*
12. H. Burkhardt and W.N. Cottingham, *Ann. Phys.* **56**, 453 (1970).
13. S. Wandzura and F. Wilczek, *Phys. Lett.* **B72**, 195 (1977).
14. D. Drechsel and L. Tiator, *J. Phys.* **G18**, 449 (1992).
15. A. Airapetian *et al.*, *Phys. Rev. Lett.* **90**, 092002-1 (2003).
16. A. Fantoni for the HERMES Collaboration, *Eur. Phys. Journal* **A17**, 385 (2003).
17. N. Bianchi, A. Fantoni and S. Liuti, *Phys. Rev.* **D69**, 014505 (2004).
18. A. Fantoni, *These proceedings.*
19. J. Yun *et al.*, *Phys. Rev.* **C67**, 055204 (2003).
20. M. Amarian *et al.*, *Phys. Rev. Lett.* **92**, 022301 (2004).
21. Z.-E. Meziani, *These proceedings.*

SPIN STRUCTURE OF THE NUCLEON AND ASPECTS OF DUALITY

Z. -E. MEZIANI

Department of Physics, Temple University
1900 N. 13 St.
Philadelphia, PA, 19122 USA
E-mail: meziani@temple.edu

In a quantitative analysis of the neutron spin structure first moment in terms of its "twist expansion", the size of higher twists (twist-4) is found to be surprisingly small at $Q^2 = 1$ GeV^2 lending support to the concept of "global duality". The value of d_2 at $Q^2 = 1$ GeV^2 tends to suggest significant higher than twist-3 contribution (twist-5 and beyond). More precision data in the range $1 \leq Q^2 \leq 5$ GeV^2 are needed for a definitive conclusion since the lattice prediction are at odds with the measured data at $Q^2 = 4.8$ GeV^2.

1. Introduction

One of the fascinating aspects of nucleon structure, known as "quark-hadron duality", is an observation made in the early 70's by Bloom and Gilman[1] while investigating the spin-independent nucleon response in deep inelastic lepton scattering (DIS). These authors found that, in the scaling regime at large momentum transfers, this response is well described by an average over the resonances structure at lower momentum transfers. Subsequently, in an attempt to explain this observation within the framework of Quantum Chromodynamics (QCD), De Rújula, Georgi and Politzer [2] used the Operator Product Expansion (OPE) method to suggest a possible link between the average over the resonances response and the DIS scaling response. While many studies were performed on the spin-independent response functions of the nucleon [3], a renewed interest has emerged in testing this "duality" behavior in the spin-dependent response functions [4,5,6]. In principle with a deeper understanding of QCD and its confinement properties we should be able to predict the observed behavior in either case. With the OPE method one has the opportunity to test the validity of our expansion at low momentum transfers by extracting the higher twist contributions and investigating the convergence and breakdown of such an expansion. Compared to the spin-independent response of the nucleon, the study the spin-dependent response offers a new variety of matrix elements of operators which describe quark-quark and quark-gluon interactions. Since to apply the OPE method integrations of the structure functions over the full excitation spectrum including the elastic scattering response is necessary, thus we shall be examining only aspects of "global duality" not "local duality" [6,3].

In the last 25 years a large amount of spin-dependent data has been collected in the DIS regime to evaluate the first moment of g_1 in order to test for example the Ellis-Jaffe and Bjorken sum rules [8,10,11,12,13,14,15,16]. Data on the g_2 structure function were collected to investigate the Burkhardt-Cottingham sum rule, quark-quark and quark-gluon correlations within the nucleon [8]. More recently g_1 and g_2 data on the neutron were obtained at Jefferson Lab at lower momentum transfers ($Q^2 \leq 1$ GeV^2) dominated by the resonance region [9]. These data combined with the world neutron data allowed for a comprehensive investigation of the Q^2 evolution from the perturbative regime to the confinement regime. In this paper we explore "global duality" by gauging its validity using the size of the extracted higher twists contributions in Γ_1 and in d_2 using the method of the OPE.

2. First moment of g_1 and size of higher twists

Using the operator product expansion (OPE) in QCD we expand the lowest moment of $\Gamma_1^n(Q^2)$ of the spin structure function g_1 of the nucleon in inverse powers of Q^2 ,

$$\Gamma_1^n(Q^2) \equiv \int_0^1 dx \; g_1^n(x, Q^2) \; = \; \sum_{\tau=2,4,\cdots} \frac{\mu_\tau^n(Q^2)}{Q^{\tau-2}} \tag{1}$$

with the coefficients μ_τ^n related to nucleon matrix elements of operators of twist $\leq \tau$. Here twist is defined as the mass dimension minus the spin of an operator, and $x = Q^2/2M\nu$ is the Bjorken x variable, with M the neutron mass. For each twist the Q^2 dependence in μ_τ^n can be calculated perturbatively as a series in α_s. Note that the application of the OPE requires summation over all hadronic final states, including the elastic at $x = 1$.

The leading-twist (twist-2) component, μ_2^n, is determined by matrix elements of the axial vector operator $\bar{\psi}\gamma_\mu\gamma_5\psi$, summed over various quark flavors. It can be decomposed into flavor triplet (g_A), octet (a_8) and singlet

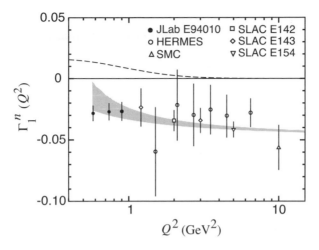

Figure 1. Q^2 dependence of Γ_1^n from various experiments. The error bars are a quadratic sum of statistical and systematic uncertainties. The twist-2 contribution is given by the band with $\Delta\Sigma = 0.35$, and its width represents the uncertainty in α_s. The elastic contribution is indicated by the dashed curve.

($\Delta\Sigma$) axial charges,

$$\mu_2^n(Q^2) = C_{ns}(Q^2)\left(-\frac{1}{12}g_A + \frac{1}{36}a_8\right) + C_s(Q^2)\frac{1}{9}\Delta\Sigma , \qquad (2)$$

where C_{ns} and C_s are the nonsinglet and singlet Wilson coefficients [17]. The nonsinglet triplet axial charge is obtained from neutron β-decay, $g_A = 1.2670(35)$ [18], while the octet axial charge is extracted from hyperon weak decay matrix elements assuming SU(3) flavor symmetry, $a_8 = 0.579(25)$ [18]. In order to factorize all of the Q^2 dependence into the Wilson coefficients, we use the renormalization group invariant definition of the matrix element of the singlet axial current, $\Delta\Sigma \equiv \Delta\Sigma(Q^2 = \infty)$.

The leading twist contribution to Γ_1^n, namely $\mu_2^n(Q^2)$, is shown in Fig. 1 along with the world data using $\Delta\Sigma = 0.35 \pm 0.08$. The value of $\Delta\Sigma$ is obtained by fitting the world data for $Q^2 \geq 5$ GeV^2 where the higher twist contributions are small. The band is a result of the uncertainty we have in α_s when evaluating the Wilson coefficients. Comparing the band with the world data suggests that the higher twists contributions are small.

The higher-twist contribution to $\Gamma_1^n(Q^2)$ is obtained by subtracting the leading-twist term from the total,

$$\Delta\Gamma_1^n(Q^2) \equiv \Gamma_1^n(Q^2) - \mu_2^n(Q^2) = \frac{\mu_4^n(Q^2)}{Q^2} + \frac{\mu_6^n(Q^2)}{Q^4} + O\left(\frac{1}{Q^6}\right). \quad (3)$$

A linear fit to the world data shows a small slope indicating that $\mu_4^n(Q^2)$ is negligible. Deviations from the linear fit indicate that higher order terms do not become significant unless the momentum transfer is below $Q^2 \leq 0.6$ GeV^2. The coefficient of the $1/Q^2$ term contains a twist-2 contribution, a_2^n, and a twist-3 term, d_2^n, in addition to the genuine twist-4 component, f_2^n [19,20,21],

$$\mu_4^n = \frac{1}{9}M^2\left(a_2^n + 4d_2^n + 4f_2^n\right) . \tag{4}$$

Interaction terms appear through the twist-3 and twist-4 contributions.

The coefficient d_2^n is given by the twist-3 part of the x^2-weighted moment of a particular combination of the measured g_1 and g_2 structure functions, and at large Q^2, corresponds to a matrix element of an operator which involves both quark and gluon fields [22].

$$d_2(Q^2) = \int_0^1 x^2\left[2g_1(x,Q^2) + 3g_2(x,Q^2)\right]dx \tag{5}$$

The twist-4 contribution, f_2 can also be expressed in terms of the structure functions:

$$f_2(Q^2) = \frac{1}{2}\int_0^1 dx\, x^2\left(7g_1(x,Q^2) + 12g_2(x,Q^2) - 9g_3(x,Q^2)\right), \tag{6}$$

where g_3 is the 3rd spin structure function, which has never been measured so far but can be accessed in the future by an asymmetry measurement of unpolarized lepton scattering off a longitudinally polarized target.

Presently f_2 has been extracted using our world knowledge of a_2 and d_2 through Eqs. 3 and 4. Values of f_2 obtained around $Q^2 = 1$ GeV^2 are found to be rather small in both the neutron and the proton lending support to the concept of "global duality" in the spin structure of the nucleon. Details of these extractions are described in Ref.[23,24]

3. Twist -3 and higher twist contributions in d_2

The quantity d_2 can be determined from a direct measurement of g_1 and g_2, it is useful to determine its size as Q^2 varies from large to small values. This allows us to gauge the size of higher twists in d_2 and extract the twist-3 matrix element. In fact this matrix element coincides with d_2 only when higher order terms are small, perhaps above $Q^2 = 1$ GeV^2. The quantity d_2 contains the elastic contribution wich becomes negligible only at Q^2 of few GeV^2 (as it shown in Fig. 2b), but can be a substantial part of the integral at low momentum transfers. As we saw in Eq. 4 d_2 enters as a higher twist term proportional to $1/Q^2$ in the expression of Γ_1. It is useful to see whether $\Delta\Gamma$ is small due of subtle cancelations of different higher twist contributions or in fact each individual twist is separatelly small.

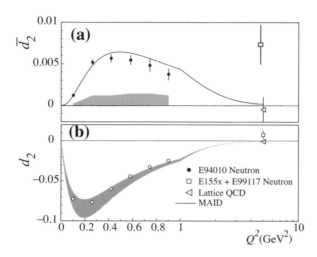

Figure 2. Measured values of $\bar{d}_2$ (a) and d_2 (b) are plotted versus the momentum transer Q^2. In the bottom panel the elastic contribution is added in the integral defining d_2. See the text for explanations. Note the change of vertical scale between the top and bottom panel.

In Fig. 2a, we show the world data of $\bar{d}_2(Q^2)$ (without the elastic contribution) at several values of Q^2 for the neutron. The results of JLab E94-010 are the solid circles and the grey band represents their corresponding systematic uncertainty. The solid line is the MAID calculation [25] which includes only the resonance contributions in the evaluation of the spin structure functions g_1 and g_2. The combined neutron result from SLAC E155X [8] and JLab E99-117 [10] around $Q^2 = 5$ GeV^2 is also shown (open square). The lattice prediction [26] at $Q^2 = 5$ GeV^2 for the neutron d_2 reduced matrix element is negative but close to zero. We point out that many nucleon models not shown in this figure predict a negative or zero value at large Q^2 where the elastic contribution is negligible.

At moderate Q^2 the data show a positive $\bar{d}_2^n$, and indicate a slow decrease with Q^2. Because MAID does not contain the DIS contribution it is unrealistic to use it as a guide to the behavior of d_2 between $Q^2 = 1$ GeV^2 and $Q^2 = 5$ GeV^2. Especially that the SLAC/JLab datum, in contrast with the lattice calculation, shows also value of $\bar{d}_2^n$ of equivalent size as the moderate momentum transfer data but, unfortunately, with still a large error bar.

When the nucleon elastic contribution is added to obtain d_2^n (Fig. 2b the situation changes dramatically, now the value of d_2 (open circles) is dominated by the elastic contribution. If one examines the data across the measured Q^2 range, we find that at large Q^2 the quantity d_2 is not of similar size to that of the moderate Q^2 region, and its sign changes from one region to the other. Data are lacking in the range $Q^2 = 1$-5 GeV^2 and thus, making it speculative to identify the measured d_2 with the twist-3 matrix element. Knowledge of the Q^2 dependence in this range is critical. The moderate Q^2 region is perhaps not suitable to be analyzed in terms of OPE due to the large size of higher twists beyond twist-3. Thus it is of paramount importance to see what the measurement of d_2 above $Q^2 = 1$ GeV^2 will show. The conclusion will depend very much on the size of the inelastic contribution below the resonances. This contribution as we mentioned before is not included in the MAID calculations. At this stage the comparison of the data and the lattice calculations to determine the size higher twists, that is twist-5 and beyond, is premature.

4. Conclusion

The size of higher twists in the OPE analysis of Γ_1^n is small thus confirming the notion of "global duality". The lack of g_2 data in the resonance region, and thus d_2, does not allow to gauge the size of the twist-5 and higher order contributions to d_2. Neutron and proton g_2 data, in the resonance region, with Q^2 ranging from 1 to 5 GeV^2 will be critical to extract with confidence the twist-3 part of d_2. The lattice QCD calculation of the twist-3 matrix element (contained in d_2) and the present world data are at odds around $Q^2 = 5$ GeV^2. At present the Q^2 dependence from 1 to 5 GeV^2 cannot be investigated, but this situation will change with the approval of JLab proposal [27] and ultimately with the 12 GeV beam energy upgrade at Jefferson Lab [28].

Acknowledgments

I thank the organizers for their invitation to speak in this well organized and focused workshop. This Work is partially supported by the Department of Energy grant DE-FG02-94ER40844.

References

1. E. D.Bloom and F. J. Gilman, *Phys. Rev. Lett.* **25**, (1970) 1140 *Phys. Rev.* **D4**, (1971) 2901.
2. A. De Rjula, H. Georgi, and H.D. Politzer, *Ann. Phys. (N.Y.)* **103**, (1977) 315.
3. W. Melnitchouk, R. Ent and C. Keppel, *Phys. Rept.* **406**, 127 (2005).
4. A. Airapetian *et al.* [HERMES Collaboration], *Phys. Rev. Lett.* **90**, 092002 (2003).
5. X. Ji and W. Melnitchouk, *Phys. Rev.* **D56**, R1 (1997).
6. N. Bianchi, A. Fantoni and S. Liuti, *Phys. Rev.* **D69**, 014505 (2004).
7. B.W. Filippone and X.D. Ji, *Adv. Nucl. Phys.* **26**, 1 (2001).
8. P.L. Anthony *et al.* [E155 Collaboration], [arXiv:hep-ex/0204028].
9. M. Amarian *et al.* [JLab E94-010 Collaboration], *Phys. Rev. Lett.* **89**, 242301 (2002), *ibid.* **92**, 022301 (2004).
10. X. Zheng *et al.* [Jefferson Lab Hall A Collaboration], *Phys. Rev. Lett.* **92**, 012004 (2003); *Phys. Rev.* **C70**, 065207 (2004).
11. P.L. Anthony *et al.* [E142 Collaboration], *Phys. Rev.* **D54**, 6620 (1996).
12. K. Abe *et al.* [E143 Collaboration], *Phys. Rev.* **D58**, 112003 (1998).
13. K. Abe *et al.* [E154 Collaboration], *Phys. Rev. Lett.* **79**, 26 (1997).
14. K. Abe *et al.* [E155 Collaboration], *Phys. Lett.* **B493**, 19 (2000).
15. A. Airapetian *et al.* [HERMES Collaboration], *Eur. Phys. J.* **C26**, 527 (2003).
16. B. Adeva *et al.* [SMC Collaboration], *Phys. Rev.* **D58**, 112001 (1998).
17. S.A. Larin, T. van Ritbergen and J.A.M. Vermaseren, *Phys. Lett.* **B404**, 153 (1997); S.A. Larin, *Phys. Lett.* **B334**, 192 (1994).
18. K. Hagiwara *et al.* [Particle Data Group Collaboration], *Phys. Rev.* **D66**, 010001 (2002).
19. E.V. Shuryak and A.I. Vainshtein, *Nucl. Phys.* **B201**, 141 (1982).
20. E. Stein, P. Gornicki, L. Mankiewicz and A. Schäfer, *Phys. Lett.* **B353**, 107 (1995).
21. X. Ji, arXiv:hep-ph/9510362.
22. X. Ji and P. Unrau, *Phys. Lett.* **B333**, 228 (1994); X. Ji, *Phys. Lett.* **B309**, 187 (1993).
23. Z.E. Meziani *et al.*, *Phys. Lett.* **B613**, 148 (2005).
24. M. Osipenko *et al.*, *Phys. Lett.* **B609**, 259 (2005).
25. D. Drechsel, S.S. Kamalov, and L. Tiator, *Phys. Rev.* **D63**, 114010 (2001).
26. Göckeler *et al.*, *Phys. Rev.* **D63**, 074506 (2001).
27. S. Choi, X. Jiang and Z.-E. Meziani spokespersons, *JLab proposal* **PR03-107**.
28. *The Science Driving the 12 GeV Upgrade at CEBAF*, http://www.jlab.org/div_dept/physics_division/GeV.html.

Duality and Confinement

QUARK MODELS OF DUALITY
IN ELECTRON AND NEUTRINO SCATTERING

W. MELNITCHOUK

Jefferson Lab, 12000 Jefferson Avenue, Newport News, VA 23606, USA
E-mail: wmelnitc@jlab.org

Results of recent analyses of electromagnetic structure functions in the resonance region suggest that duality-violating higher twists are small above $Q^2 \sim 1$ GeV^2. We analyze the systematics of local duality within a quark model framework for various modes of spin-flavor symmetry breaking. On the basis of these models we discuss expectations for the workings of duality in neutrino scattering.

1. Introduction

Hadronic cross sections averaged over appropriate energy intervals often exhibit the scaling behavior expected from perturbative QCD calculations in terms of quark and gluon degrees of freedom. A remarkable example of such quark-hadron duality is seen in inclusive electron-nucleon scattering, first noted by Bloom and Gilman.[1] Here the inclusive nucleon structure function measured in the region dominated by low-mass nucleon resonance excitations is observed to follow a global scaling function describing the high energy region, to which the resonance structure function averages. In recent years high-precision data from Jefferson Lab, HERMES, and elsewhere on spin-averaged[2] and spin-dependent[3,4] structure functions has enabled the global and local aspects of this "Bloom-Gilman (BG) duality" to be quantified, including its flavor, spin and nuclear medium dependence.[5]

Before the advent of QCD, BG duality was interpreted in the context of finite-energy sum rules, in analogy with the s and t channel duality in hadron-hadron scattering. In QCD, BG duality can be reformulated in the language of the operator product expansion (OPE), where for large Q^2 the moments of structure functions are expanded in inverse powers of the hard momentum scale Q^2, with the expansion coefficients related to matrix elements of operators of a specific twist. The leading twist (twist-2) term corresponds to scattering from free quarks, and is responsible for the scaling of structure functions (modulo perturbative α_s corrections). The higher twist terms involve multi-quark and mixed quark-gluon operators, and contain information on long-range, nonperturbative correlations between partons. The approximate independence on Q^2 of the moments is then naturally attributed to the dominance of the twist-2 term, and suppression (or cancellation) of the higher twist contributions.

In the context of global analyses of parton distributions, higher twist effects are often seen as unwelcome complications. On the other hand, higher twists contain valuable information on the nonperturbative structure of the nucleon, and can therefore provide important insights into quark confinement.

2. Higher twist matrix elements

An example of fundamental information on the structure of the nucleon which can be extracted from higher twist matrix elements are the so-called "color polarizabilities" of the nucleon. These describe how the background color electric and magnetic fields respond to the spin of the nucleon, and are defined as

$$\chi_E \, \vec{S} = \frac{1}{2M^2} \langle N| \, \vec{j}_a \times \vec{E}_a \, |N\rangle \, , \qquad \chi_B \, \vec{S} = \frac{1}{2M^2} \langle N| \, j_a^0 \, \vec{B}_a \, |N\rangle \, , \quad (1)$$

respectively, where $j_a^\mu = -g\bar{\psi}\gamma^\mu t_a \psi$ is the quark current, t_a are color SU(3) matrices, and $\vec{E}_a$ and $\vec{B}_a$ are the color electric and magnetic fields. The color polarizabilities can be expressed in terms of the twist-3 matrix element d_2, which is related to the x^2-weighted moment of the transverse g_2 structure function, and the twist-4 matrix element f_2, defined as the matrix element of the operator $\bar{\psi}\widetilde{G}^{\mu\nu}\gamma_\nu \, \psi_q$, where $\widetilde{G}^{\mu\nu}$ is the dual gluon field strength tensor.

Recently the world data on the proton and neutron (in practice ^{3}He) spin dependent g_1 structure functions have been reanalyzed[6,7] in order to consistently extract the lowest moment $\Gamma_1 = \int dx g_1$. The order $1/Q^2$ correction to Γ_1 is directly proportional to f_2, once target mass effects and the g_2 contribution are subtracted. From the values of f_2 obtained in the global analyses,[6,7] and the results for $d_2^{p,n}$ from measurements of g_2, one finds for the color polarizabilities in the proton and neutron:

$$\chi_E^p = 0.026 \pm 0.028 \, , \qquad \chi_B^p = -0.013 \mp 0.014 \, , \qquad (2)$$

$$\chi_E^n = 0.033 \pm 0.029 \, , \qquad \chi_B^n = -0.001 \pm 0.016 \, , \qquad (3)$$

where the error includes statistical and (the more dominant) systematic uncertainties, as well as from the $x \to 0$ extrapolation and uncertainty in α_s at low Q^2. These results indicate that both the color electric and magnetic polarizabilities in the proton and neutron are relatively small, with the central values of the color electric polarizabilities being positive, and the color magnetic zero or slightly negative.

The small values of the higher-twist corrections suggest that the long-range, nonperturbative interactions between quarks and gluons in the nucleon are not as dominant at $Q^2 > 1$ GeV^2 as one may have expected.[8] For the polarized neutron moment,[7] they may even play a minor role down to $Q^2 \approx 0.5$ GeV^2. This means that there are strong cancellations between nucleon resonances resulting in the dominance of the leading twist contribution to the moments. To see how such cancellations can take place, we examine a simple model in which the resonance transitions can be evaluated exactly and the degree to which duality holds quantified.

3. Local duality in a simple quark model

Duality for structure function moments can be analyzed in terms of the OPE, however no such simple interpretation exists for the x dependence of the functions themselves, or for integrals over restricted regions of x – known as "local duality". To understand the emergence of a scaling function out of resonances, one must address the question of how coherent contributions ("squares of sums of quark charges") can yield results consistent with incoherent scattering ("sums of squares of quark charges").

Close and Isgur[9] elucidated this problem by establishing the conditions necessary for duality to occur within the spin-flavor symmetric quark model. They found that for duality to hold at least one complete set of even and odd parity resonances must be summed over. Table 1 gives the relative strengths of the contributions to the proton and neutron spin averaged and spin dependent structure functions from the $N \to N^*$ transition matrix elements, for the lowest even parity $\mathbf{56^+}$ and odd parity $\mathbf{70^-}$ representations of SU(6). The coefficients λ and ρ denote the relative strengths of the symmetric and antisymmetric contributions of the SU(6) ground state wave function,

The SU(6) limit corresponds to $\lambda = \rho$. Summing over all of the states in the $\mathbf{56^+}$ and $\mathbf{70^-}$ multiplets then gives rise to a neutron to proton ratio $R^{np} \equiv F_1^{en}/F_1^{ep} = 2/3$, and polarization asymmetries $A_1^p \equiv g_1^{ep}/F_1^{ep} = 5/9$ and $g_1^{en}/F_1^{en} = 0$, just as in the quark-parton model in which the structure functions are calculated in terms of partonic (rather than N^*) degrees of freedom. In particular, for the case of g_1^{en} one sees that the cancellations

Table 1. Relative strengths of electromagnetic and weak $N \to N^*$ transitions in the quark model.[9,10]

	$^2 8^{[+]}$	$^4 10^{[+]}$	$^2 8^{[-]}$	$^4 8^{[-]}$	$^2 10^{[-]}$	total
F_1^{ep}	$9\rho^2$	$8\lambda^2$	$9\rho^2$	0	λ^2	$18\rho^2 + 9\lambda^2$
F_1^{en}	$\frac{1}{4}(3\rho + \lambda)^2$	$8\lambda^2$	$\frac{1}{4}(3\rho - \lambda)^2$	$4\lambda^2$	λ^2	$\frac{9}{2}\rho^2 + \frac{27}{2}\lambda^2$
g_1^{ep}	$9\rho^2$	$-4\lambda^2$	$9\rho^2$	0	λ^2	$18\rho^2 - 3\lambda^2$
g_1^{en}	$\frac{1}{4}(3\rho + \lambda)^2$	$-4\lambda^2$	$\frac{1}{4}(3\rho - \lambda)^2$	$-2\lambda^2$	λ^2	$\frac{9}{2}\rho^2 - \frac{9}{2}\lambda^2$
$F_1^{\nu p}$	0	$24\lambda^2$	0	0	$3\lambda^2$	$27\lambda^2$
$F_1^{\nu n}$	$\frac{1}{4}(9\rho + \lambda)^2$	$8\lambda^2$	$\frac{1}{4}(9\rho - \lambda)^2$	$4\lambda^2$	λ^2	$\frac{81}{2}\rho^2 + \frac{27}{2}\lambda^2$
$g_1^{\nu p}$	0	$-12\lambda^2$	0	0	$3\lambda^2$	$-9\lambda^2$
$g_1^{\nu n}$	$\frac{1}{4}(9\rho + \lambda)^2$	$-4\lambda^2$	$\frac{1}{4}(9\rho - \lambda)^2$	$-2\lambda^2$	λ^2	$\frac{81}{2}\rho^2 - \frac{9}{2}\lambda^2$

leading to duality already occur *within* each multiplet.

The SU(6) predictions for the structure functions hold approximately at $x \sim 1/3$, however significant deviations are observed at larger x. One can identify several mechanisms of symmetry breaking which lead to results for the large-x structure function that are consistent in the quark and hadron representations. Among these, a suppression of the symmetric (λ) configuration at large x gives rise to a suppressed d quark distribution relative to u, which in turn leads to the famous neutron to proton ratio[11] $R^{np} \to 1/4$. This is natural if one considers that if the mass difference between the nucleon and Δ is attributed to spin dependent forces, the energy associated with the symmetric part of the wave function will be larger than that of the antisymmetric component.

On the other hand, at large Q^2 one expects transition form factors, which determine the large-x behavior of structure functions, to be constrained by perturbative QCD, which predicts that photons predominantly couple to quarks with the same helicity as the nucleon. In this case the helicity-3/2 cross section will be suppressed relative to the helicity-1/2 cross section, leading to $A_1 \to 1$ for both protons and neutrons, and that the neutron to proton ratio $R^{np} \to 3/7$ — exactly as obtained in the parton level calculation on the basis of perturbative QCD.[12] Whether the $x \to 1$ behavior of structure functions follows the λ-suppression scenario or is governed by helicity conservation will be addressed experimentally at Jefferson Lab, where the 12 GeV energy upgrade will allow definitive tests of the properties of structure functions at large x.

4. Neutrino scattering

The prospect of high-intensity neutrino beams at Fermilab,[13] as well as in Japan and Europe, offers a valuable complement to the study of duality and resonance transitions. In particular, neutrino-induced reactions can provide important consistency checks on the validity of duality. Such studies will also be crucial for neutrino oscillation experiments, whose interpretation will rely heavily on understanding neutrino–nucleon interactions at low Q^2.

The main difference between electron and neutrino scattering reactions can be most easily understood considering specific resonance transitions. While a neutrino beam can convert a neutron into a proton, it cannot convert a proton into a neutron, for example (and *vice versa* for an antineutrino beam). Similarly, there are dramatic differences between inelastic production rates in the Δ resonance region[14] — because of charge conservation, only transitions to isospin-3/2 states from the proton are allowed.

A particularly relevant measurement would be of the ratio of neutron to proton neutrino structure functions at large x. Here, similar valence quark dynamics as in charged lepton scattering are probed, but with different sensitivity to quark flavors. At the hadronic level, quark model studies reveal rather distinct patterns of resonance transitions to the lowest-lying positive and negative parity SU(6) multiplets.[9,10,14]

The contributions of the $N \to N^*$ transition matrix elements to the neutrino F_1 and g_1 structure functions of the proton and neutron in the SU(6) quark model are listed in Table 1. Summation over the $N \to N^*$ transitions (for the case of equal symmetric and antisymmetric contributions to the wave function, $\lambda = \rho$) yields the expected SU(6) quark-parton model results, providing an explicit confirmation of duality. On the other hand, some modes of spin-flavor symmetry breaking ($\lambda \neq \rho$) yield neutrino structure function ratios which at the parton level are in obvious conflict with those obtained from electroproduction.[10] Neutrino structure function data can therefore provide valuable checks on the appearance of duality and its consistency between electromagnetic and weak probes.

Recently a study of duality in neutrino scattering has been undertaken[15] using phenomenologically determined weak transition form factors for both $J = 3/2$ and $J = 1/2$ resonances.[16] Currently there are only rudimentary data on resonances beyond the Δ region, and with more accurate data expected, a precise comparison will be possible in the near future.

5. Outlook

The discussion of duality in the simple quark model provides a clear illustration of how parton model results can be replicated by explicit sums over nucleon resonances. An important challenge in the future will be to consistently include the effects of both resonances and the nonresonant background in the same framework.

Future experimental exploration of duality will focus on determining its flavor, spin and target dependence, as well as its workings when probed by the weak interaction, in neutrino–nucleon scattering. Another important avenue will be to explore duality in semi-inclusive reactions. Confirmation of duality here would open the way to an enormously rich semi-inclusive program in the preasymptotic regime, allowing unprecedented access to the spin and flavor distributions of the nucleon, especially at large x.

Acknowledgments

This work was supported by the U.S. Deptartment of Energy contract DE-AC05-84ER40150, under which the Southeastern Universities Research Association operates the Thomas Jefferson National Accelerator Facility.

References

1. E. D. Bloom and F. J. Gilman, *Phys. Rev. Lett.* **25**, 1140 (1970).
2. I. Niculescu et al., *Phys. Rev. Lett.* **85**, 1186 (2000).
3. A. Airapetian et al., *Phys. Rev. Lett.* **90**, 092002 (2003).
4. N. Liyanage, *AIP Conf. Proc.* **747**, 118 (2005).
5. W. Melnitchouk, R. Ent and C. E. Keppel, *Phys. Rept.* **406**, 127 (2005).
6. M. Osipenko et al., *Phys. Lett.* **B609**, 259 (2005).
7. Z.-E. Meziani et al. *Phys. Lett.* **B613**, 148 (2005).
8. N. Bianchi, A. Fantoni and S. Liuti, *Phys. Rev.* **D69**, 014505 (2004).
9. F. E. Close and N. Isgur, *Phys. Lett.* **B509**, 81 (2001).
10. F. E. Close and W. Melnitchouk, *Phys. Rev.* **C68**, 035210 (2003).
11. W. Melnitchouk and A. W. Thomas, *Phys. Lett.* **B377**, 11 (1996).
12. G. R. Farrar and D. Jackson, *Phys. Rev. Lett* **35**, 1416 (1975).
13. D. Drakoulakos et al. [MINERνA Collaboration], arXiv:hep–ex/0405002.
14. F. E. Close and F. J. Gilman, *Phys. Rev.* **D 7**, 2258 (1972).
15. O. Lalakulich, W. Melnitchouk and E. A. Paschos, in preparation.
16. O. Lalakulich and E. A. Paschos, *Phys. Rev.* **D71**, 074003 (2005).

RECENT PREDICTIONS FROM THE STATISTICAL PARTON DISTRIBUTIONS

J. SOFFER

Centre de Physique Théorique,
*UMR 6207 *CNRS-Luminy Case 907,*
13288 Marseille Cedex 9, France
E-mail: soffer@cpt.univ-mrs.fr

A statistical model for the parton distributions in the nucleon has proven its efficiency in the global analysis of unpolarized and polarized deep-inelastic scattering data. This statistical approach involves only a few free parameters (*eight*) and has several characteristic features which will be recalled. Predictions for various QCD processes involving leptons and hadrons, are compared with recent experimental data from DESY, SLAC, FNAL, RHIC and Jefferson Lab. These new tests are very satisfactory and we will also discuss the prospect of this physical framework.

Deep-inelastic scattering (DIS) of leptons on hadrons has been extensively studied, over the last twenty years or so, both theoretically and experimentally, to extract the polarized parton distributions of the nucleon. As it is well known, the unpolarized light quarks (u, d) distributions are fairly well determined. Moreover, the data exhibit a clear evidence for a flavor-asymmetric light sea, *i.e.* $\bar{d} > \bar{u}$, and large uncertainties still persist for the gluon (G) and for the heavy quarks (s, c) distributions. The corresponding polarized gluon and s quark distributions $(\Delta G, \Delta s)$ are badly constrained and we just begin to uncover a flavor asymmetry, for the corresponding polarized light sea, namely $\Delta \bar{u} \neq \Delta \bar{d}$. Whereas the signs of the polarized light quarks distributions are essentially well established, $\Delta u > 0$ and $\Delta d < 0$, this is not the case for $\Delta \bar{u}$ and $\Delta \bar{d}$. Here we briefly recall how we construct a complete set of polarized parton distributions, namely for all flavor quarks q_i, antiquarks $\bar{q}_i$, and gluon G. Our motivation is to use the statistical approach [1,2] to build up the distributions q_i, Δq_i, $\bar{q}_i$, $\Delta \bar{q}_i$, G

---

*UMR 6207 is Unité Mixte de Recherche du CNRS and of Universités Aix-Marseille I and Aix-Marseille II, and of Université du Sud Toulon-Var, Laboratoire affilié à la FRUMAM and ΔG, in terms of a very small number of free parameters. A flavor separation for the unpolarized and polarized light sea quarks is automatically achieved in a way dictated by our approach.

The existence of the correlation, broader shape higher first moment, suggested by the Pauli principle, has inspired the introduction of Fermi-Dirac functions for the quark distributions and of Bose-Einstein functions for the gluon [3]. After many years of research, we recently proposed [1], at the input scale $Q_0^2 = 4 \text{GeV}^2$

$$xu^+(x, Q_0^2) = \frac{A X_{0u}^+ x^b}{\exp[(x - X_{0u}^+)/\bar{x}] + 1} + \frac{\tilde{A} x^{\tilde{b}}}{\exp(x/\bar{x}) + 1}, \tag{1}$$

$$x\bar{u}^-(x, Q_0^2) = \frac{\bar{A}(X_{0u}^+)^{-1} x^{2b}}{\exp[(x + X_{0u}^+)/\bar{x}] + 1} + \frac{\tilde{A} x^{\tilde{b}}}{\exp(x/\bar{x}) + 1}, \tag{2}$$

$$xG(x, Q_0^2) = \frac{A_G x^{\tilde{b}+1}}{\exp(x/\bar{x}) - 1}, \tag{3}$$

and similar expressions for the other light quarks (u^-, d^+ and d^-) and their antiparticles ($\bar{u}^-, \bar{d}^+$ and $\bar{d}^-$). It is reasonable to assume that for very small x, $xG(x, Q_0^2)$ has the same behavior as the diffractive contribution of quarks and antiquarks (second term in Eqs. (1) and (2)). We also assume $\Delta G(x, Q_0^2) = 0$ and the strange quark distributions $s(x, Q_0^2)$ and $\Delta s(x, Q_0^2)$ are simply related [1] to $\bar{q}(x, Q_0^2)$ and $\Delta \bar{q}(x, Q_0^2)$, for $q = u, d$. A peculiar aspect of this procedure, is that it solves the problem of disentangling the q and $\bar{q}$ contribution through the relationship [1] $X_{0u}^+ + X_{0\bar{u}}^- = 0$, and the corresponding one for the other light quarks and their antiparticles. It allows to get the $\bar{q}(x)$ and $\Delta \bar{q}(x)$ distributions from the ones for $q(x)$ and $\Delta q(x)$. By performing a next-to-leading order (NLO) QCD evolution of these parton distributions, we were able to obtain in Ref. [1], a good description of a large set of very precise data on the following unpolarized and polarized DIS structure functions $F_2^{p,d,n}(x, Q^2), xF_3^{\nu N}(x, Q^2)$ and $g_1^{p,d,n}(x, Q^2)$, in a broad range of x and Q^2, in correspondance with the following *eight* free parameters :

$$X_{0u}^+ = 0.46128, \ X_{0u}^- = 0.29766, \ X_{0d}^- = 0.30174, \ X_{0d}^+ = 0.22775, \tag{4}$$

$$\bar{x} = 0.09907, \ b = 0.40962, \ \tilde{b} = -0.25347, \ \tilde{A} = 0.08318, \tag{5}$$

and three additional parameters, which are fixed by normalization conditions

$$A = 1.74938, \ \bar{A} = 1.90801, \ A_G = 14.27535. \tag{6}$$

Given the numerical values of the parameters, crucial tests will be provided by measuring flavor and spin asymmetries for antiquarks and in particular, we expect [1]

$$\Delta \bar{u}(x) > 0 > \Delta \bar{d}(x), \tag{7}$$

$$\Delta \bar{u}(x) - \Delta \bar{d}(x) \simeq \bar{d}(x) - \bar{u}(x) > 0. \tag{8}$$

The inequality $\bar{d}(x) - \bar{u}(x) > 0$ has the right sign to agree with the defect in the Gottfried sum rule. The statistical model gives the value 0.176, for the Bjorken integral, in excellent agreement with the QCD prediction 0.182 ± 0.005 and with the world data $0.176 \pm 0.003 \pm 0.07$ [4]. We also note that, if Eq. (8) above is satisfied, it means that the antiquark polarization contributes to it and in the statistical model this contribution is 0.022, which is not negligible. For illustration, we show in Fig. 1 some predictions of the statistical approach, with recent unpolarized DIS results [5]. The description of the data is indeed very satisfactory, in spite of its limited accuracy, mainly in the large x region at high Q^2. This kinematic region requires additional data, for providing crucial tests to our approach.

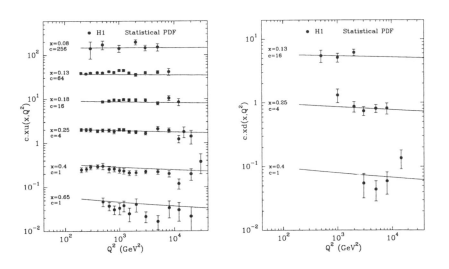

Figure 1. Statistical quark distributions $c \cdot xu(x, Q^2)$, $c \cdot xd(x, Q^2)$ as a function of Q^2 for fixed x bins. Data from H1, Ref. [5] (Taken from Ref. [6]) .

In Fig. 2, we show the remarkable agreement of the statistical model with recent data on polarized parton distributions. We compare our predictions with an attempt from Hermes[7,8] to isolate the different quark and antiquark helicity distributions. The poor quality of the data does not allow to conclude on the signs of $\Delta\bar{u}(x)$ and $\Delta\bar{d}(x)$, which will have to wait for a higher precision experiment. From recent accurate Jefferson Lab data [9] on $\Delta u(x)/u(x)$ and $\Delta d(x)/d(x)$, we note the behavior near $x = 1$, a typical property of the statistical approach, at variance with predictions of the current literature. The fact that $\Delta u(x)$ is more concentrated in the higher x region than $\Delta d(x)$, accounts for the change of sign of $g_1^n(x)$, which becomes positive for $x > 0.5$, a behavior first observed at Jefferson Lab [10].

A very important source of information for $\bar{q}(x)$ distributions comes from

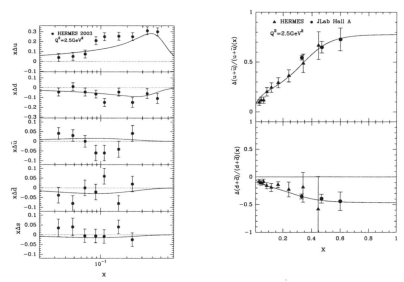

Figure 2. On the left, polarized quark and antiquark distributions, as a function of x for $Q^2 = 2.5\text{GeV}^2$. Data from Hermes Ref. [7] and the curves are predictions from the statistical approach. On the right, results for $\Delta(u+\bar{u})/(u+\bar{u})(x)$ and $\Delta(d+\bar{d})/(d+\bar{d})(x)$ from Refs. [8,9], compared to the statistical model predictions (Taken from Ref. [6]).

Drell-Yan dilepton processes, whose cross sections are proportional to the products of $q(x)$ and $\bar{q}(x)$ distributions. In Fig. 3 we compare successfully our predictions and the data for Drell-Yan pair production at Tevatron.

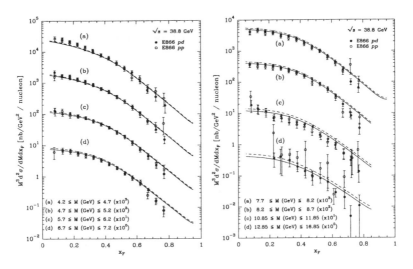

Figure 3. Drell-Yan cross sections per nucleon at $\sqrt{s} = 38.8$GeV for pp and pd as a function of x_F for selected M bins. Solid curve pp, dashed curve pd. Experimental data from FNAL E866, Ref. [11] (Taken from Ref. [6]).

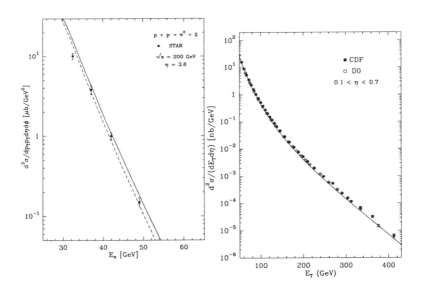

Figure 4. On the left, inclusive π^0 production in pp reaction at $\sqrt{s} = 200$GeV as a function of E_π. Data from STAR, Ref. [12] (Taken from Ref. [6]). On the right, cross section for single jet production in $\bar{p}p$ at $\sqrt{s} = 1.8$TeV as a function of E_T. Data are from CDF, Ref. [13] and D0, Ref. [14] experiments (Taken from Ref. [6]).

Finally, in Fig. 4 our NLO QCD predictions are in perfect agreement with the cross sections data for single jet production at Tevatron, in the central rapidity region, and for inclusive π^0 production at RHIC, at large rapidity.

We also look forward to the future of the spin programme at RHIC, which will provide new spin asymmetries data, to be confronted with the basic predictions of this physical framework.

Acknowledgments

I would like to thank the organizers of this very successful workshop on quark-hadron duality , in particular Nicola Bianchi, for the invitation and for giving me the opportunity to deliver this talk.

References

1. C. Bourrely, F. Buccella and J. Soffer, *Eur. Phys. J.* **C23**, 487 (2002). For a practical use of these PDF, see www.cpt.univ-mrs.fr/~bourrely /research/bbs-dir/bbs.html.
2. C. Bourrely, F. Buccella and J. Soffer, *Mod. Phys. Lett.* **A18**, 771 (2003).
3. C. Bourrely, F. Buccella, G. Miele, G. Migliore, J. Soffer and V. Tibullo, *Z. Phys.* **C62**, 431 (1994).
4. SLAC E155 Collaboration, P. L. Anthony *et al.*, *Phys. Lett.* **B493**, 19 (2000).
5. H1 Collaboration, C. Adloff *et al.*, *Eur. Phys. J.* **C30**, 1 (2003) [hep-ex/0304003].
6. C. Bourrely, F. Buccella and J. Soffer, *Eur. Phys. J.* **C41**, 327 (2005).
7. HERMES Collaboration, A. Airapetian *et al.*, *Phys. Rev. Lett.* **92**, 012005 (2004)
8. HERMES Collaboration, K. Ackerstaff *et al.*, *Phys. Lett.* **B464**, 123 (1999).
9. Jlab E-99-117 Collaboration, X. Zheng *et al.*, *Phys. Rev.* **C70**, 065207 (2004).
10. Jlab E-99-117 Collaboration, X. Zheng *et al.*, *Phys. Rev. Lett.* **92**, 012004 (2004).
11. FNAL E866/NuSea Collaboration, J.C. Webb *et al.*, submitted to Phys. Rev. Lett. [hep-ex/0302019]
12. STAR Collaboration, G. Rakness, contribution to the XI Int. Workshop on Deep Inelastic Scattering (DIS2003), 23-27 April 2003, St. Petersburg, Russia; S. Heppelmann, contribution to the Transversity Workshop, 6-7 October 2003, IASA, Athens, Greece; J. Adams *et al.*, Phys. Rev. Lett. **92**, 171801 (2004) [hep-ex/0310058]
13. CDF Collaboration, T. Affolder *et al.*, *Phys. Rev.* **D64**, 032001 (2001).
14. D0 Collaboration, B. Abbott *et al.*, *Phys. Rev.* **D64**, 032003 (2001).

HADRON STRUCTURE ON THE BACK OF AN ENVELOPE

A. W. THOMAS, R. D. YOUNG

Thomas Jefferson National Accelerator Facility
12000 Jefferson Ave.,
Newport News, VA 23185, USA
E-mail: awthomas@jlab.org

D. B. LEINWEBER

Special Research Center for the Subatomic Structure of Matter, and
Department of Physics, University of Adelaide,
Adelaide, SA 5005 Australia
E-mail: dleinweb@physics.adelaide.edu.au

In order to remove a little of the mysticism surrounding the issue of strangeness in the nucleon, we present simple, physically transparent estimates of both the strange magnetic moment and charge radius of the proton. Although simple, the estimates are in quite good agreement with sophisticated calculations using the latest input from lattice QCD. We further explore the possible size of systematic uncertainties associated with charge symmetry violation (CSV) in the recent precise determination of the strange magnetic moment of the proton. We find that CSV acts to increase the error estimate by 0.003 μ_N such that $G_M^s = -0.046 \pm 0.022 \ \mu_N$.

1. Introduction

The tremendous amount of experience that has been gained over the last 6 years, by studying the chiral extrapolation of lattice QCD data as a function of quark (or pion) mass, has led to very important insights into hadron structure. The two major lessons learned are that:

- The contribution of pion loops to hadron properties decreases very fast as the pion mass increases, becoming small and slowly varying for pion masses above about 500 MeV [1,2,3]. As a corollary, it follows that the effect of kaon loops is always relatively small[4] – an issue we shall return to soon in the context of strangeness form factors.
- Once the pion mass is of the order of 500 MeV or higher, **all** hadron properties are smooth, slowly varying and essentially behave like

41

the constituent quark model. The corollary to this is that if one wishes to build a constituent quark model of hadron structure, this is the mass region where it has a chance to work[5] – far from the region of rapidly varying non-analytic behaviour associated with pions near the chiral limit.

The second lesson is of particular relevance to the understanding of duality, because in this mass region ($m_\pi > 500$ MeV) the reconstruction of the valence parton distribution functions (PDFs) shows that each valence quark does indeed have a most likely momentum fraction around 1/3 [6], precisely as one would expect in a naive constituent quark picture.

One of the remarkable things that became obvious from the beginning of these studies is the fact that the relatively naive cloudy bag model[8,9] (CBM) did an astonishingly good job of describing the mass dependence of nucleon properties, whether it be the mass[7], magnetic moments[3] or moments of the PDFs[6]. This does not mean that the CBM is ideal, we see from the comparison against G_{En} [10,11], in particular, that the sharp surface of the MIT bag, upon which the CBM was built, is not such a good description of the valence quark structure, especially in the surface region[12]. However, what **does** seem to really describe the way hadron structure works is that there is a perturbative pion cloud around a core of confined valence quarks, confined in a region whose vacuum structure (the bag itself) differs from that of the QCD ground state.

The recent discovery that the chiral quark soliton model also yields the correct dependence of m_N on m_π [13] (apart from the incorrect chiral coefficient associated with the hedgehog approximation) is consistent with this interpretation, since even though one has to work extremely hard to construct the change in the vacuum structure inside the nucleon at the microscopic level, in the end it looks like a system of bound valence quarks surrounded by a perturbative pion cloud. The consequences of the change in vacuum structure inside the hadron, in terms of a contribution to $\bar{d} \neq \bar{u}$ and $\Delta\bar{u} \neq \Delta\bar{d}$ are in fact similar in both models[14,15].

Further support for this idea comes from a remarkable discovery concerning the lattice QCD data for the nucleon and the Δ in both quenched (QQCD) and full QCD (QCD). In fact, one can describe the data with a simple fitting function, $\alpha + \beta\, m_\pi^2$ plus the pion self-energy loops which give rise to the leading (LNA) and next-to-leading non-analytic (NLNA) behaviour, evaluated using a finite range regulator of dipole form (with common mass parameter $\Lambda = 0.8$ GeV). The important discovery is that

α and β (for a given baryon) are the **same** within the fitting errors (a few percent) in QQCD and QCD[2]. This is the case even though for the Δ the self-energies differ by a factor of two, with the N-Δ splitting being of order 500 MeV in QQCD and only 300 MeV in full QCD. It seems that the "core" or valence structure of these key baryons, defined by the particular value of $\Lambda = 0.8$ GeV, is essentially the same in QQCD and QCD. The implications of this for modeling hadron structure are yet to be fully investigated but once again a simple, perturbative treatment of the pion cloud contributions works exceptionally well.

One of the most impressive recent achievements of the chiral extrapolation program has been the determination of an extremely precise value for the strangeness magnetic moment, G_M^s [16]. This calculation used a combination of experimental data for the octet magnetic moments, the constraints of charge symmetry and chiral extrapolation of state of the art lattice data to obtain the ratios of the magnetic moments of either a valence u quark in the proton and Σ^+ or the valence u quark in the neutron and the Ξ^0. By reducing the demands on lattice QCD to mere ratios, it is possible to dramatically reduce the systematic errors. The result, namely $G_M^s = -0.046 \pm 0.019 \mu_N$ is an order of magnitude more precise than any current experiment [17,18,19,20,21] – a unique example in modern hadron physics. A similar analysis for the strangeness electric form factor, G_E^s, has not yet been possible, essentially because the experimental knowledge of octet baryon charge radii is nowhere near as precise as the knowledge of magnetic moments. However, our main purpose, to which we turn in the next section, is to use what we have learnt so far about hadron structure to make a "back of the envelope" estimate of both the strangeness electric and magnetic form factors. Then in the following section, we provide an estimate of the systematic uncertainty associated with charge symmetry violation in the recent precise determination of the strangeness magnetic moment of the nucleon[16].

2. Simple Model of the Strangeness Form Factors of the Proton

We note first that there is no known example where the current quark masses show up in hadron physics undressed by non-perturbative glue. Thus the cost to make an $s - \bar{s}$ pair in the proton is of order 1.0 to 1.1 GeV (twice the strange constituent quark mass). On the other hand, creating the $\bar{s}$ in a kaon and the s in a Λ costs only 0.65 GeV. (Note that the N to

$K\Sigma$ coupling is considerably smaller than that for N to $K\Lambda$ and hence in this simple discussion we ignore it.) On these grounds alone we expect the virtual transition N to $K\Lambda$ to dominate the production of strangeness in the proton.

Next we estimate the probability for finding the $K\Lambda$ configuration. This probability is inversely proportional to the excitation energy squared. We work by comparison with the N π component of the nucleon wave function, for which there is a vast body of evidence that it is about 20% [22]. Naively the transition N to N π costs 140 MeV, but with additional kinetic energy this is around 600 MeV in total. Including similar kinetic energy for the $K\Lambda$ component as well, it costs roughly twice as much as N π. Thus the $K\Lambda$ probability is of order 5%.

2.1. *Strangeness radius*

We consider first the strangeness radius of the proton based on this 5% $K\Lambda$ probability. In the CBM the radius of a Λ bag is about 1 fm, which yields a mean square radius for the strange quark around 0.5 fm^2. As an estimate of the range of variation possible, we also take the bag radius $R = 0.8$ fm with a corresponding mean square radius close to 0.36 fm^2. In order to estimate the contribution from the kaon cloud, we need to realize that in almost any chiral quark model the peak in the Goldstone boson wave function is at the confinement (bag) radius[12,24]. As long ago as 1980 this generated enormous interest in the precise measurement of G_E^n [23]. The meson field then decreases with a range between one over the energy cost of the Fock state and $1/(m_K + m_\Lambda - m_N)$. Thus for $R = 0.8$ fm we get a mean square radius for the $\bar{s}$ distribution of order 1 fm^2, while for $R = 1$ fm we get about 1.4 fm^2. Weighting the s by $-1/3$ and $\bar{s}$ by $+1/3$, we find that the mean square charge radius of strange quarks is between $(-0.36+1.0)/3$ and $(-0.5 + 1.4)/3$, that is in the range $(0.21, 0.30)$ fm^2, times the probability for finding the $K\Lambda$ configuration.

To calculate G_E^s at $Q^2 = 0.1$ GeV$^2 = 2.5$ fm^{-2}, we assume that the term $-Q^2 < r^2 > /6$ dominates and finally multiply by -3 to agree with the usual convention of removing the strange quark charge. This yields $G_E^s \in (+0.01, +0.02)$. It is definitely small and definitely positive for the very clear physical reasons that the $K\Lambda$ probability is small and that the kaon cloud extends outside the Λ. A comparison with the currently preferred fit to the existing world data[20] reveals that this estimate is in agreement at the 1 σ level, although it is nominally of opposite sign.

2.2. *Strangeness magnetic moment*

Because orbital angular momentum is quantized, the contribution to the magnetic moment from the $\bar{s}$ in the kaon cloud is much less model dependent. The Clebsch-Gordon coefficients show that in a spin-up proton the probability of a spin down (up) Λ, accompanied by a kaon with orbital angular momentum $+1$ (0), is $2/3$ (1/3). We also know the magnetic moment of the Λ and that it is dominated by the magnetic moment of the s quark. Hence the total strangeness magnetic moment of the proton is $-3 \times P_{K\Lambda} \times 2/3 \times (+0.6 + 1/3) - 3 \times P_{K\Lambda} \times 1/3 \times (-0.6 + 0)$, where the terms in brackets are, respectively, the magnetic moment of the spin down (up) Λ and the magnetic moment of the charge $+1/3$ $\bar{s}$ quark with one unit (or zero units) of orbital angular momentum. The nett result, namely $G_M^s = -0.063\ \mu_N$, is reasonably close to the best lattice QCD estimate noted above, that is $G_M^s = -0.046 \pm 0.019\mu_N$. From the point of view of this "back of the envelope" estimate, the lattice result clearly has both a natural magnitude and sign. It is very hard to see how the result could change much in either magnitude or sign unless the physical picture presented here is totally incorrect. Given the remarks concerning our present understanding of hadron structure based on experience with the study of chiral extrapolation and lattice QCD data, this seems unlikely.

3. Impact of Charge Symmetry Violation on G_M^s

In the spirit of Refs. [25,26], we use p, n, u^p etc., to denote the magnetic moment of that baryon or, in the case of a quark, to denote the valence quark sector contribution of that flavor to that baryon *if* that quark had unit charge. A valence quark sector contribution is depicted in the left-hand diagram of Fig. 1. We also denote the total contribution of u, d and s quarks in a "disconnected loop" in baryon B, depicted in the right-hand diagram of Fig. 1, as O_B. By determining O_p, the strangeness magnetic moment of the proton can be obtained by calculating the ratio of strange to non-strange loop contributions.

The magnetic moments of the octet baryons satisfy:

$$p = e_u\, u^p + e_d\, d^p + O_p\ ; \qquad n = e_d\, d^n + e_u\, u^n + O_n\ ,$$
$$\Sigma^+ = e_u\, u^{\Sigma^+} + e_s\, s^{\Sigma^+} + O_{\Sigma^+}\ ; \quad \Sigma^- = e_d\, d^{\Sigma^-} + e_s\, s^{\Sigma^-} + O_{\Sigma^-}\ ,$$
$$\Xi^0 = e_s\, s^{\Xi^0} + e_u\, u^{\Xi^0} + O_{\Xi^0}\ ; \quad \Xi^- = e_s\, s^{\Xi^-} + e_d\, u^{\Xi^-} + O_{\Xi^-}\ .$$

$$(1)$$

Having removed the charge factors from the valence quark contributions to

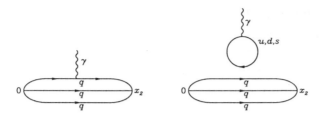

Figure 1. Diagrams illustrating the two topologically different insertions of the current within the framework of lattice QCD. In full QCD these diagrams are dressed with an arbitrary number of gluons and additional quark loops.

baryon magnetic moments, it is usually at this point that charge symmetry is invoked to provide equivalence between the doubly-represented u-quark sector of the proton u^p and the doubly-represented d-quark sector of the neutron, d^n. Similarly, u^{Σ^+} is taken to be equal to d^{Σ^-}, and u^{Ξ^0} is taken to be equal to d^{Ξ^-}. However, current quark mass differences of a few MeV and electromagnetic effects will act to violate these equalities. In these cases, charge symmetry violation (CSV) in the quark flavor being probed by the electromagnetic current is directly related to the differences observed in baryon properties.

However, indirect environmental effects are also induced through CSV. For example, even though it is the same strange quark that appears in Σ^+ and Σ^-, its contributions to the baryon moment will differ due to subtle differences in the environment of the strange quark. Similar environmental effects will provide subtle violations of $s^{\Xi^0} = s^{\Xi^-}$, $O_p = O_n$, $O_{\Sigma^+} = O_{\Sigma^-}$ and $O_{\Xi^0} = O_{\Xi^-}$.

Introducing Δ_B to denote the contribution to the magnetic moment of baryon B having its origin in CSV, Eqs. (1) take the exact forms

$$
\begin{aligned}
p &= e_u\, u^p + e_d\, d^p + O_p\;; & n &= e_d\, u^p + e_u\, d^p + O_p - \Delta_n\,, \\
\Sigma^+ &= e_u\, u^{\Sigma^+} + e_s\, s^{\Sigma^+} + O_{\Sigma^+}\;; & \Sigma^- &= e_d\, u^{\Sigma^+} + e_s\, s^{\Sigma^+} + O_{\Sigma^+} - \Delta_{\Sigma^-}\,, \\
\Xi^0 &= e_s\, s^{\Xi^0} + e_u\, u^{\Xi^0} + O_{\Xi^0}\;; & \Xi^- &= e_s\, s^{\Xi^0} + e_d\, u^{\Xi^0} + O_{\Xi^0} - \Delta_{\Xi^-}\,.
\end{aligned}
$$
$$(2)$$

While we have elected to write the right-hand expressions of Eqs. (2) in terms of quantities in the left-hand expressions and Δ_B, we note that one could have done the opposite and this will be important in quantifying Δ_B.

The total sea-quark loop contribution to the proton magnetic moment, O_p, includes sea-quark-loop contributions from u, d and s quarks (right-hand side of Fig. 1). By definition

$$O_p = \frac{2}{3}{}^\ell G_M^u - \frac{1}{3}{}^\ell G_M^d - \frac{1}{3}{}^\ell G_M^s, \tag{3}$$

$$= \frac{1}{3}{}^\ell G_M^d - \frac{1}{3}{}^\ell G_M^s - \frac{2}{3}\Delta_{loop}, \tag{4}$$

where we have introduced

$${}^\ell G_M^u = {}^\ell G_M^d - \Delta_{loop}, \tag{5}$$

with Δ_{loop} accounting for differences in the u and d sea-quark loop contributions to the proton due to direct CSV.

Introducing, in the usual fashion, the ratio of s- to d-quark loop contributions, ${}^\ell R_d^s \equiv {}^\ell G_M^s/{}^\ell G_M^d$, Eq. (4) provides

$$O_p = \frac{{}^\ell G_M^s}{3}\left(\frac{1 - {}^\ell R_d^s}{{}^\ell R_d^s}\right) - \frac{2}{3}\Delta_{loop}, \tag{6}$$

The established approach[16] centres around two equations for the strangeness magnetic moment of the nucleon, G_M^s, obtained from linear combinations of the above. Including the Δ_B terms to account for CSV, one has the exact relations

$$G_M^s = \left(\frac{{}^\ell R_d^s}{1 - {}^\ell R_d^s}\right)\left[2p + 2\Delta_{loop} + n + \Delta_n - \frac{u^p}{u^\Sigma}\left(\Sigma^+ - \Sigma^- - \Delta_{\Sigma^-}\right)\right], \tag{7}$$

$$G_M^s = \left(\frac{{}^\ell R_d^s}{1 - {}^\ell R_d^s}\right)\left[p + 2\Delta_{loop} + 2n + 2\Delta_n - \frac{u^n}{u^\Xi}\left(\Xi^0 - \Xi^- - \Delta_{\Xi^-}\right)\right]. \tag{8}$$

The ratios u^p/u^Σ and u^n/u^Ξ are ratios of valence-quark contributions to baryon magnetic moments in full QCD as depicted in the left-hand diagram of Fig. 1. The latter are determined by lattice QCD calculations and finite range regularization effective field theory techniques[16] with the results

$$\frac{u^p}{u^\Sigma} = 1.092 \pm 0.030 \quad \text{and} \quad \frac{u^n}{u^\Xi} = 1.254 \pm 0.124. \tag{9}$$

The ratio of s- and d-quark sea-quark loop contributions, ${}^\ell R_d^s \equiv G_M^s/{}^\ell G_M^d$, has been estimated conservatively[16,28] as 0.139 ± 0.042.

Tests of CSV suggest that it is typically smaller than a 1% effect in baryon properties. The structure of Eqs. (7) and (8) suggests that a good estimate of the systematic uncertainty in G_M^s would be provided by taking the CSV terms Δ_B to represent uncertainties with a magnitude of 1% of the associated baryon moment.

As discussed following Eqs. (2), the CSV corrections Δ_{Σ^-}, and Δ_{Ξ^-} could equally well have been represented on the left-hand expressions of

Eqs. (2) as Δ_{Σ^+}, and Δ_{Ξ^0}. Hence we will replace Δ_{Σ^-}, and Δ_{Ξ^-} in Eqs. (7) and (8) with Δ_Σ, and Δ_Ξ representing the average 1% uncertainties of the hyperon charge states. Our focus on O_p for strangeness in the proton, does not allow a similar symmetry for the nucleon.

Since the underlying mechanisms giving rise to CSV, as represented by Δ_B, are different for each baryon, the uncertainties are accumulated in quadrature. Focusing on Eq. (8) where the error is largest, this provides a CSV uncertainty of 0.011 μ_N to G_M^s. Given the already large error on $G_M^s = -0.046 \pm 0.019 \ \mu_N$ associated with statistical, scale determination, chiral correction and $^\ell R_d^s$ uncertainties[28], this additional CSV uncertainty has only a small effect on the final error estimate. Adding the CSV uncertainty in quadrature provides a total uncertainty of 0.022 μ_N.

4. Concluding Remarks

In the light of recent insights into hadron structure based on lattice QCD and the associated work on chiral extrapolation using a finite range regulator, we have explained how to quickly and easily estimate the strangeness electric and magnetic form factors of the proton. The resulting ranges, $G_E^s(0.1\text{GeV}^2) \in (+0.01, +0.02)$ and $G_M^s = -0.063\mu_N$ are relatively small, certainly challenging for our experimental colleagues, but consistent within 95% CL with current world data. The latter is also in remarkable agreement with the recent determination based on lattice QCD.

We also explored the size of systematic uncertainties associated with charge symmetry violation in the recent precise determination of the strange magnetic moment of the proton[16]. We find CSV acts to increase the error estimate by 0.003 μ_N such that $G_M^s = -0.046 \pm 0.022 \ \mu_N$. Hence even accounting for CSV in the approach, one still has a two-sigma signal for the sign of the strange magnetic moment of the proton.

In conclusion, this is a crucial point in the history of the study of hadron structure. For the first time we have useful guidance from non-perturbative QCD using the methods of lattice QCD and chiral extrapolation. These rigorous calculations can be given life through the sort of simple physical model described here, which nevertheless permits semi-quantitative calculation. At the same time we have new experimental capabilities to accurately measure the role of non-valence quarks in static properties which can be used to test our new found theoretical advances.

Acknowledgements

This work is supported by the Australian Research Council and by DOE contract DE-AC05-84ER40150, under which SURA operates Jefferson Laboratory.

References

1. W. Detmold, D. B. Leinweber, W. Melnitchouk, A. W. Thomas and S. V. Wright, *Pramana* **57**, 251 (2001).
2. R. D. Young, D. B. Leinweber, A. W. Thomas and S. V. Wright, *Phys. Rev.* **D66**, 094507 (2002).
3. D. B. Leinweber, D. H. Lu and A. W. Thomas, *Phys. Rev.* **D60**, 034014 (1999).
4. V. V. Flambaum *et al.*, *Phys. Rev.* **D69**, 115006 (2004).
5. I. C. Cloet, D. B. Leinweber and A. W. Thomas, *Phys. Rev.* **C65**, 062201 (2002).
6. W. Detmold, W. Melnitchouk and A. W. Thomas, *Eur. Phys. J. direct* **C3**, 13 (2001).
7. D. B. Leinweber, *et al.*, *Phys. Rev.* **D61**, 074502 (2000).
8. A. W. Thomas, *Adv. Nucl. Phys.* **13**, 1 (1984).
9. G. A. Miller, *Int. Rev. Nucl. Phys.* **1**, 189 (1984).
10. D. I. Glazier *et al.*, *Eur. Phys. J.* **A24**, 101 (2005).
11. J. M. Finn [E93038 Collaboration], *FizikaB* **13** (2004) 545.
12. D. H. Lu, A. W. Thomas and A. G. Williams, *Phys. Rev.* **C57**, 2628 (1998).
13. K. Goeke, J. Ossmann, P. Schweitzer and A. Silva, arXiv:hep-lat/0505010.
14. A.W. Schreiber, A.I. Signal and A.W. Thomas, *Phys. Rev.* **D44** (1991) 2653.
15. B. Dressler, *et al.*, arXiv:hep-ph/9809487.
16. D. B. Leinweber *et al.*, *Phys. Rev. Lett.* **94**, 212001 (2005).
17. D. T. Spayde *et al.* [SAMPLE Collaboration], *Phys. Rev. Lett.* **84**, 1106 (2000).
18. E. J. Beise, M. L. Pitt and D. T. Spayde, *Prog. Part. Nucl. Phys.* **54**, 289 (2005).
19. F. E. Maas *et al.*, *Phys. Rev. Lett.* **94**, 082001 (2005).
20. D. S. Armstrong *et al.* [G0 Collaboration], arXiv:nucl-ex/0506021.
21. K. A. Aniol *et al.* [HAPPEX Collaboration], arXiv:nucl-ex/0506011.
22. J. Speth and A. W. Thomas, *Adv. Nucl. Phys.* **24**, 83 (1997).
23. A. W. Thomas, S. Theberge and G. A. Miller, *Phys. Rev.* **D24**, 216 (1981).
24. A. W. Thomas, J. D. Ashley, D. B. Leinweber and R. D. Young, *J. Phys. Conf. Ser.* **9**, 321 (2005).
25. D. B. Leinweber, *Phys. Rev.* **D53**, 5115 (1996).
26. D. B. Leinweber and A. W. Thomas, *Phys. Rev.* **D62**, 074505 (2000).
27. I. Eschrich *et al.* [SELEX Collaboration], *Phys. Lett.* **B522**, 233 (2001).
28. D. B. Leinweber, *et al.*, *Eur. Phys. J.* **A24S2**, 79 (2005).

THE STRONG COUPLING CONSTANT AT LOW Q^2

A. DEUR

Thomas Jefferson National Accelerator Facility
12000 Jefferson Avenue, Newport News, VA 23606, USA
E-mail: deurpam@jlab.org

We extract an effective strong coupling constant using low-Q^2 data and sum rules. Its behavior is established over the full Q^2-range and is compared to calculations based on lattice QCD, Schwinger-Dyson equations and a quark model. Although the connection between all these quantities is not known yet, the results are surprisingly alike. Such a similitude may be related to quark-hadron duality.

1. The strong coupling constant

A peculiar feature of strong interaction is asymptotic freedom: quark-quark interactions grow weaker with decreasing distances. Asymptotic freedom is expressed in the vanishing of the QCD coupling constant,$\alpha_s(Q^2)$, at large Q^2. Conversely, the fact that $\alpha_s(Q^2)$, as calculated in pQCD, becomes large when $Q^2 \to \Lambda_{QCD}^2$ is often linked to quark confinement. Since it is not expected that pQCD holds at the confinement scale and since the condition $\alpha_s(Q^2) \to \infty$ when $Q \to \Lambda_{QCD}$ is far from necessary to assure confinement[1], it is interesting to study $\alpha_s(Q^2)$ in the large distance domain.

Experimentally, moments of structure functions are convenient objects to extract α_s. Among them, Γ_1^{p-n} is the simplest to use. In pQCD, it is linked to the axial charge of the nucleon, g_A, by the Bjorken sum rule:

$$\Gamma_1^{p-n} \equiv \int_0^1 dx(g_1^p(x) - g_1^n(x)) = \frac{1}{6}g_A[1 - \frac{\alpha_s}{\pi} - 3.58\left(\frac{\alpha_s}{\pi}\right)^2 \tag{1}$$

$$-20.21\left(\frac{\alpha_s}{\pi}\right)^3 - 130.0\left(\frac{\alpha_s}{\pi}\right)^4 - 893.38\left(\frac{\alpha_s}{\pi}\right)^5] + \sum_{i=2}^{\infty}\frac{\mu_{2i}^{p-n}}{Q^{2i-2}},$$

where $g_1^p(g_1^n)$ is the first spin structure function for the proton(neutron). The $\mu_t(Q^2)/Q^{t-2}$ are higher twist corrections and become important at lower Q^2. This series, usually truncated to leading twist and to 3rd order, can be used to fit experimental data and to extract α_s. The higher twists can be computed with non-perturbative models or can be extracted from data, although with limited precision at the moment[3]. This imprecise knowledge and the break down of pQCD at low Q^2 prevent a priori the extraction of α_s at low Q^2. However, an *effective* strong coupling constants was defined by Grunberg[4] in which higher twists and higher order QCD radiative corrections are incorporated. Eq. 1 becomes by definition:

$$\Gamma_1^{p-n} \equiv \frac{1}{6}g_A[1 - \frac{\alpha_{s,g_1}}{\pi}]. \tag{2}$$

This definition yields many advantages: the coupling constant is extractable at any Q^2, is well-behaved when $Q^2 \to \Lambda_{QCD}$, is not renormalization scheme (RS) dependent and is analytic when crossing quark thresholds. The price to pay for such benefits is that it becomes process-dependent (hence the subscript g_1 in Eq. 2). However, as pointed out by Brodsky *et al.*[5], effective couplings can be related to each other, at least in the pQCD domain, by "commensurate scale equations". These relate, using different Q^2 scales, observables without RS or scale ambiguity. Thus, one effective coupling constant is enough to characterize the strong interaction.

Among the possible observables available to define an effective coupling constant, Γ_1^{p-n} has unique advantages. The generalized Gerasimov-Drell-Hearn (GDH)[6,7] and Bjorken sum rules predict Γ_1^{p-n} at low and large Q^2, and Γ_1^{p-n} is experimentally known between these two domains. Hence, α_{s,g_1} can be extracted at any Q^2. In particular, it has a well defined value at $Q^2=0$. Furthermore, we will see that α_{s,g_1} might best be suited to be compared to the predictions of theories and models.

2. Experimental determination of α_{s,g_1}

A measurement of Γ_1^{p-n} at intermediate Q^2 was reported recently[8] and was used to extract α_{s,g_1}[9]. The results are shown by the triangles in Fig. 1, together with α_{s,g_1} extracted from SLAC data[10] at $Q^2=5$ GeV^2 (open square). Note that the elastic contribution is not included in Γ_1^{p-n}.

Γ_1^{p-n} is related to the generalized GDH sums:

$$\Gamma_1^{p-n} = \frac{Q^2}{16\pi^2\alpha}(GDH^p - GDH^n) \tag{3}$$

where α is the QED coupling constant. Hence, at $Q^2=0$, $\Gamma_1^{p-n} = 0$ and

$$\alpha_{s,g_1} = \pi. \tag{4}$$

At $Q^2 = 0$, the GDH sum rule implies:

$$\Gamma_1^{p-n} = \frac{Q^2}{16\pi^2\alpha}(GDH^p - GDH^n) = \frac{-Q^2}{8}(\frac{\kappa_p^2}{M_p^2} - \frac{\kappa_n^2}{M_n^2}) \tag{5}$$

where κ_p (κ_n) is the proton (neutron) anomalous magnetic moment. Combining Eq. 2 and 5, we get the derivative of α_{s,g_1} at $Q^2=0$:

$$\frac{d\alpha_{s,g_1}}{dQ^2} = \frac{3\pi}{4g_A} \times (\frac{\kappa_n^2}{M_n^2} - \frac{\kappa_p^2}{M_p^2}). \tag{6}$$

Relations 4 and 6 constrain α_{s,g_1} at low Q^2 (dashed line in Fig. 1). At large Q^2, Γ_1^{p-n} can be estimated using Eq. 1 at leading twist and α_s calculated with pQCD. α_{s,g_1} can be subsequently extracted (gray band).

These data and sum rules give $\alpha_{s,g_1}(Q^2)$ at any Q^2. A similar result is obtained using a model of Γ_1^{p-n} and Eq. 2 (dotted line). The Burkert-Ioffe[11] model is used because of its good match with data.

One can compare our result to effective coupling constants extracted using different processes. $\alpha_{s,\tau}$ was extracted from τ-decay data[12] from the OPAL experiment (inverted triangle). It is compatible with α_{s,g_1}. The Gross-Llewellyn Smith sum rule[13] (GLS) can be used to form α_{s,F_3}. The sum rule relates the number of valence quarks in the hadron, n_v, to the structure function $F_3(Q^2, x)$. At leading twist, it reads:

$$\int_0^1 F_3(Q^2, x)dx = n_v \left[1 - \frac{\alpha_s(Q^2)}{\pi} - 3.58 \left(\frac{\alpha_s(Q^2)}{\pi} \right)^2 - 20.21 \left(\frac{\alpha_s(Q^2)}{\pi} \right)^3 \right]$$

We expect $\alpha_{s,F_3} = \alpha_{s,g_1}$ at high Q^2, since the Q^2-dependence of Eq. 1 and 7 at leading twist are identical. The GLS sum was measured by the CCFR collaboration[14] and the resulting α_{s,F_3} is shown by the star symbols.

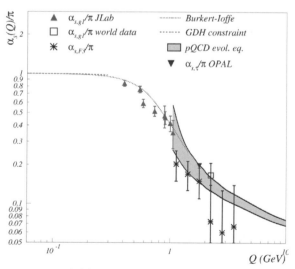

Figure 1. Extracted $\alpha_{s,g_1}(Q)/\pi$ using JLab data (up triangles), the GLS sum rule (stars), the world Γ_1^{p-n} data (open square), the Bjorken sum rule (gray band) and the Burkert-Ioffe Model. $\alpha_{s,\tau}(Q)/\pi$ from OPAL is given by the reversed triangle. The dashed line is the GDH constrain on the derivative of $\alpha_{s,g_1}/\pi$ at $Q^2=0$.

3. Comparison with theory

Just like effective coupling constants extracted experimentally, there are also many possible theory definitions for the coupling constant and, contrarily to the experimental quantities, the relations between the various definitions are not well known. Furthermore, the connection between the experimental and the theoretical quantities is not clear. Hence, the remainder of this paper is to be understood as a candid comparison of quantities *a priori* defined differently, in order to see if they share common features.

Calculations of α_s using Schwinger-Dyson equations (SDE), lattice QCD or quark models are available. Different SDE results are shown in Fig. 2. The pioneering result of Cornwall[15] is shown by the blue band in the top left panel. The more recent SDE results from Fisher *et al.*, Bloch *et al.*, Maris and Tandy, and Bhagwat *et al.* are shown in top left, top right, bottom left and bottom left panels respectively. There is a good match between the data and the result from Fisher *et al.* and a fair match with the curve from Bloch *et al.* The results from Maris-Tandy, Bhagwat *et al.* and Cornwall do not match the data. The Godfrey and Isgur curve in the top right panel of Fig. 2 is the coupling constant used in the framework of hadron spectroscopy[20]. Q^2-behavior of coupling constants can also be compared regardless of their absolute magnitudes by normalizing them to π at $Q^2 = 0$ (These curves are not shown here). The Godfrey-Isgur, Cornwall and Fisher *et al.* Q^2-behavior match well the data. The normalized curves from Maris-Tandy, Bloch *et al.* and Bhagwat *et al.* are slightly below the data (by typically one sigma) for $Q > 0.6$ GeV.

Gluon bremsstrahlung and vertex corrections contribute to the running of α_s. Modern SDE calculations include those[21] but it is *a priori* not the case for the α_s used in the one gluon exchange term of the Godfrey and Isgur quark model, or for older SDE works. If so, pQCD corrections should be added to these calculations. The effect of those corrections (*on* α_{s,g_1}) is given by the ratio of α_{s,g_1} extracted using Eq. 2 to α_{s,g_1} extracted using Eq. 1 at leading twist. For both Eq. 1 and 2, Γ_1^{p-n} is given by a model[11]. Since model and data agree well, no strong model dependence is introduced. The difference between results using Eq. 1 up to 4th and 5th order is taken as the uncertainty due to the truncation of the pQCD series. The resulting α_s are shown in the bottom right panel of Fig. 2.

Finally, we can compare lattice QCD data to our results. Many lattice results are available and are in general consistent. We chose to compare with the results of Furui and Nakajima[22], see bottom left panel in Fig. 2.

They match well the data. The lowest Q^2 point is afflicted by finite size effect and should be ignored.

The match between our data and the various calculations might be surprising since these quantities are defined differently. We can try to understand this fact. Choosing Γ_1^{p-n} minimizes the rôle of resonances, in particular it fully cancels the Δ_{1232} contribution which usually dominates the moments at low Q^2. By furthermore excluding the elastic contribution, we obtain a quantity for which coherent reactions (elastic and resonances) are suppressed and we are back to a DIS-like case in which the interpretation is straightforward. One can also possibly invoke the phenomenon of quark-hadron duality to explain why the extraction of α_{s,g_1}, using a formalism developed for DIS[12], seems to also work at lower Q^2.

4. Conclusion

We have extracted, using JLab data at low Q^2 together with sum rules, an effective strong coupling constant at any Q^2. A striking feature is its loss of Q^2-dependence at low Q^2. We compared our result to SDE and lattice QCD calculations and to a coupling constant used in a quark model. Despite the unclear relation between these various coupling constants, data and calculations match in most cases, especially for relative Q^2-dependences. This could be linked to quark-hadron duality.

Acknowledgments

This work is supported by the U.S. Department of Energy (DOE). The Southeastern Universities Research Association (SURA) operates the Thomas Jefferson National Accelerator Facility for the DOE under contract DE-AC05-84ER40150.

References

1. See e.g. Y. L. Dokshitzer, hep-ph/9812252.
2. J.D. Bjorken, *Phys. Rev.* **148**, 1467 (1966).
3. J-P. Chen, A. Deur, Z-E. Meziani, nucl-ex/0509007.
4. G. Grunberg, *Phys. Lett.* **B95** 70 (1980), *Phys. Rev.* **D29** 2315 (1984), *Phys. Rev.* **D40**, 680 (1989).
5. S.J. Brodsky and H.J. Lu, *Phys. Rev.* **D51** 3652 (1995); S.J. Brodsky, G.T. Gabadadze, A.L. Kataev and H.J. Lu, *Phys. Lett.* **B372** 133 (1996); S.J. Brodsky, hep-ph/0310289.
6. S.D. Drell and A.C. Hearn, *Phys. Rev. Lett.* **16**, 908 (1966). S. Gerasimov, *Sov. J. Nucl. Phys.* **2**, 430 (1966).
7. X. Ji and J. Osborne, *J. Phys.* **G27** 127 (2001).
8. A. Deur *et al.*, *Phys. Rev. Lett.* **93** 212001 (2004).
9. A. Deur *et al.*, hep-ph/0509113.

56

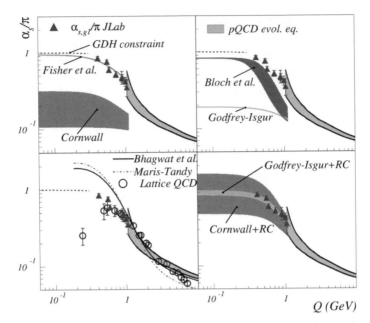

Figure 2. α_{s,g_1} extracted from JLab data and sum rules compared to various calculations: top left panel: SDE calculations from Fisher *et al.* and Cornwall; top right panel: Bloch *et al.* (SDE) and Godfrey-Isgur (quark model); bottom left: Furui and Nakajima (lattice QCD), Maris-Tandy (SDE) and Bhagwat *et al.* (SDE); bottom right: the Godfrey-Isgur and Cornwall results with pQCD radiative corrections added.

10. K. Abe *et al.*, *Phys. Rev. Lett.* **79** 26 (1997); P.L. Anthony *et al.*, *Phys. Lett.* **B493** 19 (2000), *Phys. Rev.* **D67** 055008 (2003).
11. V.D. Burkert and B.L. Ioffe, *Phys. Lett.* **B296**, 223 (1992), *J. Exp. Theor. Phys.* **78**, 619 (1994).
12. S.J. Brodsky *et al.*, *Phys. Rev.* **D67** 055008 (2003).
13. D.J. Gross and C.H. Llewellyn Smith, *Nucl. Phys* **B14** 337 (1969).
14. J.H. Kim *et al.*, *Phys. Rev. Lett.* **81** 3595 (1998).
15. J.M. Cornwall, *Phys. Rev.* **D26** 1453 (1982).
16. C.S. Fischer and R. Alkofer, *Phys. Lett.* **B536** 177 (2002); C.S. Fischer, R. Alkofer and H. Reinhardt, *Phys. Rev.* **D65** 125006 (2002); R. Alkofer, C.S. Fischer and L. von Smekal, *Acta Phys. Slov.* **52** 191 (2002).
17. J.C.R. Bloch, *Phys. Rev.* **D66** 034032 (2002).
18. P. Maris and P.C. Tandy, *Phys. Rev.* **C60** 055214 (1999).
19. Bhagwat *et al.*, *Phys. Rev.* **C68** 015203 (2003).
20. S. Godfrey and N. Isgur, *Phys, Rev.* **D32** 189 (1985).
21. We thank P. Tandy for pointing it out to us.
22. S. Furui, H. Nakajima, hep-lat/ 0410038; S. Furui and H. Nakajima, *Phys. Rev.* **D70** 094504 (2004).

SPIN DUALITY ON THE NEUTRON (^{3}HE)

P. SOLVIGNON

FOR THE JEFFERSON LAB HALL A COLLABORATION

Temple University, Department of Physics
1900 N. 13th Street,
Philadelphia, PA 19122, USA
E-mail: solvigno@jlab.org

Thomas Jefferson National Accelerator Facility experiment E01-012 measured the ^{3}He spin structure functions and virtual photon asymmetries in the resonance region in the range $1.0 < Q^2 < 4.0$ (GeV/c)^2. Our data, when compared with existing deep inelastic scattering data, can be used to test quark-hadron duality in g_1 and A_1 for ^{3}He and the neutron. Preliminary results for $A_1^{^3\text{He}}$ are presented, as well as an overview of the experimental and theoretical developments.

1. Motivation

Quark-Hadron Duality was first observed in 1970 by Bloom and Gilman[1] for the spin independent structure function F_2. The authors noticed that the nucleon resonances average on the F_2 scaling curve. Recent data[2,3] on the proton unpolarized and polarized structure functions measured in the resonance region indicate the onset of duality at momentum transfers as low as 0.5 and 1.6 (GeV/c)^2 respectively.

Substantial efforts are ongoing to investigate quark-hadron duality in polarized structure functions both experimentally and theoretically. Carlson and Mukhopadhyay[4] showed within perturbative QCD that structure functions in the resonance region fall with increasing Q^2 at the same rate as in the deep inelastic scattering (DIS) region. In the high x region, the photon is more likely to interact with the quark having the same helicity as the nucleon. This implies that both g_1 and F_1 behave as x approaches 1 and $(1-x)^3$ as $x \to 1$. A_1 is expected[5] to tend to 1 as $x \to 1$ in the scaling region. Carlson and Mukhopadhyay, considering resonant contributions and non-resonant background, predict the same behavior in the resonance region at large enough momentum transfer.

Recently, Close and Melnitchouk[6] studied three different conditions of

SU(6) symmetry breaking in the resonance region under which predictions of the structure functions at large x lead to the same result as the parton model. They examined the conditions in which certain resonances are removed from the summation (suppression of spin-$\frac{3}{2}$, suppression of helicity-$\frac{3}{2}$ and suppression of symmetric wave function), and found that each scenario predicts $A_1^{n,p} \to 1$ as $x \to 1$.

Many more theorists are working on this exciting phenomenon[7,8]

Now that precise spin structure data[9] in the DIS region are available at large x, data in the resonance region are needed in order to test polarized duality. The goal of experiment E01-012 was to provide such data on the neutron (^{3}He) in the moderate Q^2 region up to $Q^2 = 4.0$ (GeV/c)^2 where duality is expected to hold.

2. The experiment

E01-012 ran successfully in January-February 2003 at Jefferson Lab in Hall A. It was an inclusive measurement of longitudinally polarized electrons scattering off a longitudinally or transversely polarized ^{3}He target[10]. Asymmetries and cross section differences were measured in order to extract the spin structure function g_1 and the virtual photon asymmetry A_1 in the resonance region:

$$g_1 = \frac{MQ^2\nu}{4\alpha_e^2}\frac{E}{E'}\frac{1}{E+E'}\left[\Delta\sigma_\parallel + \tan(\frac{\theta}{2})\Delta\sigma_\perp\right] \qquad (1)$$

$$A_1 = \frac{A_\parallel}{D(1+\eta\xi)} - \frac{\eta A_\perp}{d(1+\eta\xi)} \qquad (2)$$

where $A_\parallel$ ($A_\perp$) is the parallel (perpendicular) asymmetry corrected for data acquisition deadtime, beam charge asymmetry, target and beam polarizations and nitrogen dilution. $\Delta\sigma_{\parallel(\perp)} = 2A_{\parallel(\perp)}\sigma_0$ with σ_0 the unpolarized cross section, and η, ξ, D and d are kinematic factors (see for example [9]). D and d depend on $R(x,Q^2)$ for which we will use recent data[11] from Jefferson Lab experiment E94-110. However our data allows a direct extraction of g_1 (and g_2) without the need of external input.

The incident electron beam was produced by a strained GaAs cathode illuminated with a polarized diode laser. The helicity of the longitudinally polarized electrons is inverted at 30Hz when passing through a pockell cell. The beam polarization was measured with a Möller polarimeter to be between 70 and 85% depending on the incident energy.

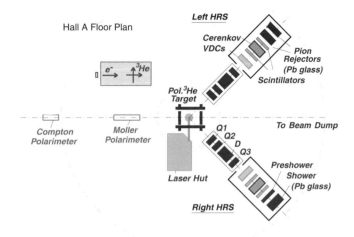

Figure 1.   Hall A floor plan. The electron beam is coming from the left.

The polarized ^{3}He target is based on spin exchange between optically pumped rubidium and ^{3}He. A small amount of nitrogen is added as buffering gas. With two sets of Helmhotz coils and two sets of laser optics, longitudinal and transverse polarizations can both be achieved. The polarization of the target was monitored approximately every 4 hours by two independent polarimetries[10]: NMR and EPR. From the preliminary analysis, the average target polarization was $(37.0 \pm 1.5)\%$.

The two almost identical Hall A High Resolution Spectrometers[10] were used in a symmetric configuration (Fig. 1) in order to double the statistics and provide a systematic cross check. The detector package contains two double planes of vertical drift chambers (VDC) for particle tracking, two planes of scintillators used to trigger the data acquisition, and a gas Čerenkov counter plus two layer electromagnetic calorimeter for particle identification. The PID detectors allowed us to reduce the pion contamination by a factor better than 10^4 while keeping the electron efficiency above 99%.

3. Preliminary results

$A_1^{^3\mathrm{He}}$ was extracted from our data. For this analysis, a constant value of $R(x,Q^2)=0.18$ was used. A complete analysis will be done later in order to evaluate $R(x, Q^2)$ for ^{3}He at our kinematics. The nitrogen dilution was determined from data taken with an identical cell filled with nitrogen. Radiative corrections have been applied. Fig. 2 shows $A_1^{3\text{He}}$ at four different Q^2 values. The position of the $\Delta(1232)$ resonance is indicated for each subset of data. The most noticeable feature is the negative contribution of the $\Delta(1232)$ resonance at low Q^2. It has been argued[4,6] that quark-hadron duality should not work in the Δ region at low Q^2. However, at Q^2 above 2.0 $(\text{GeV/c})^2$, the dominant negative bump at the location of $\Delta(1232)$ seems to vanish. Furthermore the results from these higher Q^2 settings show that $A_1^{3\text{He}}$ become positive with increasing x, following the same trend as the DIS world data[9,12]. $A_1^{3\text{He}}$ from the two highest Q^2 settings agree well with each other showing no strong Q^2-dependence.

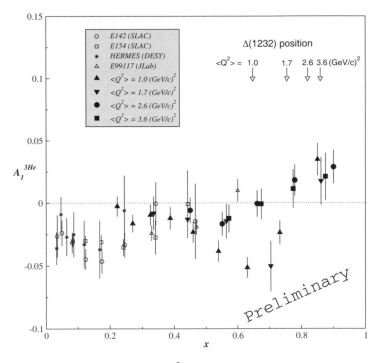

Figure 2.   Preliminary result of $A_1^{3\text{He}}$. The error bars are statistical only.

Results of the spin asymmetry A_2 and the spin structure functions g_1 and g_2 for ^{3}He were also presented at this workshop.

The polarized ^{3}He target was used in this experiment as an effective neutron target. Because of the dominant S-state of ^{3}He where the two protons have their spins anti-aligned, we can expect neutron spin structure functions to show similar behavior as observed for ^{3}He structure functions here. Work is ongoing to extract the neutron results from the ^{3}He results.

4. Summary

Experiment E01-012 provides spin structure data in the resonance region for the neutron (^{3}He) for $1.0 < Q^2 < 4.0$ (GeV/c)^2 and $0.2 < x < 0.90$. At $x < 0.60$, where DIS data are available, these data will allow a test of quark-hadron duality for neutron and ^{3}He spin structure functions. Preliminary results show that $A_1^{^3\text{He}}$ in the resonance region follows the same behavior as $A_1^{^3\text{He}}$ measured in the DIS region. These data will also be used to extract moments[13] of the structure functions, for example: the extended GDH sum, the d_2 matrix element and the Burkhardt-Cottingham sum rule.

5. Acknowledgments

This work was supported by DOE contract DE-AC05-84ER40150 under which the Southeastern Universities Research Association (SURA) operates the Thomas Jefferson National Accelerator Facility.

References

1. E.D. Bloom and F.J. Gilman, *Phys. Rev. Lett.* **25**, 1140 (1970).
2. I. Niculescu *et al.*, *Phys. Rev. Lett.* **85**, 1182 (2000), *ibid* **85**, 1186 (2000).
3. A. Airapetian *et al.*, *Phys. Rev. Lett.* **90**, 092002 (2003).
4. C.E. Carlson and N.C. Mukhopadhyay, *Phys. Rev.* **D58**, 094029 (1998).
5. G.R.Farrar and D.R. Jackson, *Phys. Rev. Lett.* **35**, 1416 (1975).
6. F.E. Close and W. Melnitchouk, *Phys. Rev.* **C68**, 035210 (2003).
7. N. Bianchi, A. Fantoni and S. Liuti, *Phys. Rev.* **D69**, 014505 (2004).
8. A. Fantoni *These proceedings.*
9. X. Zheng *et al.*, *Phys. Rev. Lett.* **92**, 012004 (2004), *Phys. Rev.* **C70**, 065207 (2004).
10. J. Alcorn *et al.*, *Nucl. Instrum. Meth.* **A522**, 294 (2004).
11. Y. Liang *et al.*, [nucl-ex/0410027], *Phys. Rev. Lett* submitted.
12. P.L. Anthony *et al.*, *Phys. Rev.* **D54**, 6620 (1996); S. Incerti, Ph.D. thesis, Université Blaise Pascal, Clermont-Ferrand, France, 1998; K. Ackerstaff *et al.*, *Phys. Lett.* **B464**, 123 (1999)
13. Z.-E. Meziani *These proceedings.*

LOCAL DUALITY IN SPIN STRUCTURE FUNCTIONS g_1^p AND g_1^d

YELENA PROK

Massachusetts Institute of Technology
stationed at Thomas Jefferson National Accelerator Facility
12000 Jefferson Ave, MS 16B/12
Newport News, Virginia 12606 E-mail: yprok@jlab.org

FOR THE CLAS COLLABORATION

Inclusive double spin asymmetries obtained by scattering polarized electrons off polarized protons and deuterons have been analyzed to address the issue of quark-hadron duality in the polarized spin structure functions g_1^p and g_1^d. A polarized electron beam, solid polarized NH$_3$ and ND$_3$ targets and the CEBAF Large Acceptance Spectrometer (CLAS) in Hall B were used to collect the data. The resulting g_1^p and g_1^d were averaged over the nucleon resonance energy region ($M < W < 2.00$ GeV), and three lowest lying resonances individually for tests of global and local duality.

1. Duality in spin structure function g_1

An observation that the hadronic and partonic degrees of freedom can sometimes both be successfully used to describe the structure of hadrons is called quark-hadron duality. This phenomenon was discovered experimentally by Bloom and Gilman[1], who observed that the spin averaged structure function $F_2(\nu, Q^2)$ measured in the resonance region was on average equivalent to the deep inelastic one, if averaged over the variable $w' = (2M\nu + M^2)/Q^2$. Quark-hadron duality can be quantified by considering partial moments of the resonance structure functions at fixed Q^2:

$$\int_{\xi_{min}}^{\xi_{max}} F_2(\xi, Q^2) d\xi, \quad \xi = \frac{2x}{(1 + (1 + 4M^2x^2/Q^2))^{-0.5}} \tag{1}$$

These moments are compared to the integrals of the 'scaling' structure functions in the same region of ξ at the same Q^2. The polarized parton distribution (PDF) based fits and the phenomenological fits to the DIS scattering data at high Q^2 with the ξ values corresponding to those of the resonant data are used as the 'scaling' functions. The equivalence of 'resonant' and 'scaling' moments is referred to as 'global' duality, if the integration is taken over the classical resonance region $(0.94 < W < 2.00$ GeV). If the average is taken over individual resonance regions, the 'local' duality is invoked. The duality is tested by forming a ratio of integrals of $g_1^{p,d}$ over the resonance region, or a particular resonance, to the integral of the deep-inelastic $g_1^{p,d}$ over an equivalent region in ξ:

$$I = \frac{\int_{\xi_{min}}^{\xi_{max}} g_1^{p,d}(data)(Q^2,\xi)d\xi}{\int_{\xi_{min}}^{\xi_{max}} g_1^{p,d}(fit_{DIS})(Q^2,\xi)d\xi} \tag{2}$$

Duality is realized when this ratio is consistent with unity.

2. Measurements and Analysis

g_1 was extracted from measurements of the double spin asymmetry $A_{\parallel}$ in inclusive ep scattering:

$$g_1 = \frac{F_1}{1+\gamma^2}[A_{\parallel}/D + (\gamma - \eta)A_2], \tag{3}$$

where F_1 is the unpolarized structure function, A_2 is the virtual photon asymmetry, and γ, D and η are kinematic factors. F_1 and A_2 are calculated using a parametrization of the world data, and $A_{\parallel}$ is measured. The spin asymmetry for ep scattering is given by:

$$A_{\parallel} = \frac{N_- - N_+}{N_- + N_+} \frac{C_N}{f P_b P_t f_{RC}} + A_{RC}, \tag{4}$$

where $N_-(N_+)$ is the number of scattered electrons normalized to the incident charge with negative (positive) beam helicity, f is the dilution factor needed to correct for the electrons scattering off the unpolarized background, f_{RC} and A_{RC} correct for radiative effects, and C_N is the correction factor associated with polarized ^{15}N nuclei in the target. $A_{\parallel}$ was measured by scattering polarized electrons off polarized nucleons using a cryogenic solid polarized target and CLAS in Hall B.

The longitudinally polarized electrons were produced by a strained $GaAs$ electron source with a typical beam polarization of $\sim 70\%$. Two solid polarized targets were used: ^{15}ND$_3$ for polarized deuterons and ^{15}NH$_3$ for polarized protons. The targets were polarized using the method of Dynamic Nuclear Polarization, with the typical polarization of $70-90\%$ for protons, and $10-35\%$ for deuterons. Besides the polarized targets, three unpolarized targets (^{12}C, ^{15}N, liquid ^{4}He) were used for background measurements. The scattered electrons were identified using the CLAS package[2], consisting of drift chambers, Cherenkov detector, time-of-flight counters and electro-magnetic calorimeters. Data were taken with beam energies of 1.6, 2.4, 4.2 and 5.7 GeV, covering a kinematic range of of $0.05 < Q^2 < 4.5$ GeV^2 and $0.8 < W < 3.0$ GeV.

3. Results

Fig. 1 shows g_1^p plotted vs ξ in Q^2 bins, compared with the next-to-leading order twist-2 PDF predictions[3],[4] and the DIS SLAC fit[5]. Each version of

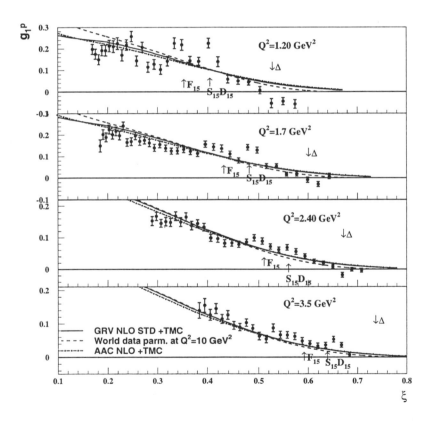

Figure 1. g_1^p plotted vs ξ in Q^2 bins, compared with the 2 PDF models and the DIS fit at Q^2=10 GeV^2. The PDF models are corrected for the target mass effects as described in the text.

the pQCD calculation has been corrected for the target mass effects[6], taking into account the fact that the measurements were taken at a low Q^2. Both PDF curves agree well on average with the data points, with exception of the resonance regions. The data points in the second and third resonances tend to lie above the fits, and the region of the $\Delta(W = 1232)$ resonance is 'below' the fits at $Q^2 < 3.0$ GeV^2. The DIS fit is consistent with the PDF predictions, beginning to deviate at low values of ξ.

In order to test the global and local duality for $g_1^{p,d}$, the ratio in equation 2 was evaluated for the whole resonance region ($1.07 < W < 2$ GeV), and three low-mass resonance regions: $1.12 < W < 1.38$ GeV ($\Delta P_{33}(1232)$), $1.38 < W < 1.58$ GeV ($D_{13}(1520), S_{11}(1535)$), and $1.58 < W < 1.8$ GeV ($F_{15}(1680)$ and others). The numerator of the above equation is displayed as data points, weighted by Q^2, with the curves showing the Q^2 weighted denominator, evaluated by integrating over the structure function g_1^p predicted by the PDFs and the DIS data fit at $Q^2=10$ GeV^2. An effect of adding the elastic contribution to the numerator was also tested, with the elastic contribution evaluated from the elastic form factors [7].

Results of this study are shown in Fig. 2. Without the inclusion of the elastic constribution, the onset of global ($1.07 < W < 2.00$ GeV) duality for the proton and deuteron is observed when Q^2 values exceed 1.2 GeV^2, while including the elastic constribution delays it until the higher values of Q^2. The local duality does not appear to hold for the proton in the first and second resonance regions, with the first region showing negative asymmetry due to the $\Delta(1232)$ constribution, and the second region having a large positive asymmetry due to the negative parity resonances S_{11} and D_{13}. In case of a deuteron, the local duality appears to hold in the second and third resonance regions.

References

1. E. Bloom and F. Gilman, *Phys. Rev.* **D4**, 2901 (1971).
2. B.A. Mecking et al., *Nucl.Instr.Meth* **503/3**, 513 (2003).
3. M. Gluck, E. Reya, M. Stratmann and W. Vogelsang *Phys. Rev.* **D63**, 014505 (2001).
4. M. Hirai *et al.*, *Phys. Rev.* **D69**, 054021 (2004).
5. K. Abe, *Phys.Rev.* **D58**, 112003 (1998).
6. J. Blumlein and A. Tkabladze, *Nucl. Phys.* **B553**, 427 (1999).
7. P. Bosted, *Phys. Rev.* **C51**, 409 (1995).

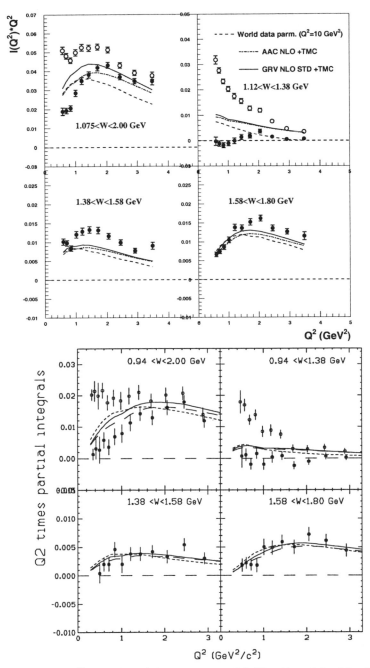

Figure 2. Top: The Q^2 evolution of the numerator of I (Eq. 2), weighted by Q^2, in the entire resonance region, and 3 individual regions, for the proton. Bottom: Same for the deuteron. Empty circles correspond to the inclusion of the elastic contribution. The deviations from unity smaller than 10% are not considered significant in this study, as there is systematic uncertainty in the data on the order of 6-7 %, and a similar uncertainty in the DIS fit.

SPIN-FLAVOR DECOMPOSITION IN POLARIZED SEMI-INCLUSIVE DEEP INELASTIC SCATTERING EXPERIMENTS AT JEFFERSON LAB

XIAODONG JIANG [*]

*Department of Physics and Astronomy,
Rutgers, the State University of New Jersey,
136 Frelinghuysen Road, Piscataway, NJ 08854 USA.
E-mail: jiang@jlab.org*

A Jefferson Lab experiment proposal was discussed in this talk. The experiment is designed to measure the beam-target double-spin asymmetries A_{1n}^h in semi-inclusive deep-inelastic $\vec{n}(\vec{e}, e'\pi^+)X$ and $\vec{n}(\vec{e}, e'\pi^-)X$ reactions on a longitudinally polarized ^{3}He target. In addition to A_{1n}^h, the flavor non-singlet combination $A_{1n}^{\pi^+ - \pi^-}$, in which the gluons do not contribute, will be determined with high precision to extract $\Delta d_v(x)$ independent of the knowledge of the fragmentation functions. The data will also impose strong constraints on quark and gluon polarizations through a global NLO QCD fit.

1. Introduction

Polarized semi-inclusive deep-inelastic scattering (SIDIS) experiments can be used to study the spin-flavor structure of the nucleon, as has been demonstrated first by the SMC experiment [1]. Recently, the HERMES experiment published results of a leading order spin-flavor decomposition from polarized proton and deuteron SIDIS asymmetry data, and for the first time extracted the $\bar{u}, \bar{d}$ and $s = \bar{s}$ sea quark polarizations [2]. The HERMES "purity" method of spin-flavor decomposition relies on the assumption that the quark fragmentation process and the experimental phase spaces are well-understood such that a LUND model based Monte Carlo simulation can reliably reproduce the probability correlations between the detected hadrons and the struck quarks [2]. The accuracies on the knowledge of the fragmentation process played a crucial role in extracting the polarized parton distributions in this method.

---

*Work supported by the U.S. National Science Foundation grant NSF-PHY-03-54871.

As an alternate method, Christova and Leader pointed out [3] that if the flavor non-singlet combination $A_1^{\pi^+ - \pi^-}$ is measured, the quark polarization Δu_v, Δd_v and $\Delta \bar{u} - \Delta \bar{d}$ can be extracted at leading order without the complication of fragmentation functions. In fact, information on the valence quark polarizations will be well preserved at any QCD order in $A_1^{\pi^+ - \pi^-}$ since gluons do not contribute to this flavor non-singlet observable. At leading order, assuming isospin symmetry and charge conjugation, the fragmentation functions cancel exactly in $A_1^{\pi^+ - \pi^-}$ and the s-quarks do not contribute, so that:

$$A_{1p}^{\pi^+ - \pi^-} \equiv \frac{\Delta \sigma_p^{\pi^+} - \Delta \sigma_p^{\pi^-}}{\sigma_p^{\pi^+} - \sigma_p^{\pi^-}} = \frac{A_{1p}^{\pi^+} - A_{1p}^{\pi^-} \cdot \sigma_p^{\pi^-}/\sigma_p^{\pi^+}}{1 - \sigma_p^{\pi^-}/\sigma_p^{\pi^+}} = \frac{4\Delta u_v - \Delta d_v}{4u_v - d_v},$$

$$A_{1n}^{\pi^+ - \pi^-} \equiv \frac{\Delta \sigma_n^{\pi^+} - \Delta \sigma_n^{\pi^-}}{\sigma_n^{\pi^+} - \sigma_n^{\pi^-}} = \frac{A_{1n}^{\pi^+} - A_{1n}^{\pi^-} \cdot \sigma_n^{\pi^-}/\sigma_n^{\pi^+}}{1 - \sigma_n^{\pi^-}/\sigma_n^{\pi^+}} = \frac{4\Delta d_v - \Delta u_v}{4d_v - u_v}. \quad (1)$$

Thus, measurements of $A_1^{\pi^+ - \pi^-}$ on the proton and the neutron can determine Δu_v and Δd_v. On the other hand, another non-singlet quantity is constrained by the inclusive data:

$$g_1^p(x, Q^2) - g_1^n(x, Q^2) = \frac{1}{6} \left[(\Delta u + \Delta \bar{u}) - (\Delta d + \Delta \bar{d}) \right] |_{LO}. \quad (2)$$

If one is only interested in flavor non-singlet quantities, such as $\Delta \bar{u} - \Delta \bar{d}$, the goal of SIDIS experiments is reduced to obtaining information on $\Delta u_v - \Delta d_v$, and the polarized sea asymmetry can be extracted at leading order :

$$(\Delta \bar{u} - \Delta \bar{d})|_{LO} = 3(g_1^p - g_1^n)|_{LO} - \frac{1}{2}(\Delta u_v - \Delta d_v)|_{LO}. \quad (3)$$

At next-to-leading order, QCD global fits of data from both inclusive and semi-inclusive reactions have become the state of the art [4]. One expects that the next generation of polarized parton distribution functions will take advantage of improvements on both quality and volume of SIDIS data from HERMES, COMPASS and Jefferson Lab. Although there are data available and experiments planned [6] with polarized proton and deuteron targets, there has been a lack of attention in obtaining SIDIS data on the neutron from a polarized ^{3}He target.

2. A proposal of polarized ^{3}He SIDIS at Jefferson Lab

An experiment proposal [7] has been developed recently at Jefferson Lab Hall A to provide high statistics neutron SIDIS data using a polarized ^{3}He target. The plan is to measure the double-spin asymmetries A_{1n}^h in $\vec{n}(\vec{e}, e'h)X$

reactions ($h = \pi^+$ and π^-, with K^+ and K^- as by-products) with a 6 GeV polarized electron beam on a longitudinally polarized ^{3}He target at a luminosity of 10^{36} sec^{-1}cm^{-2}. The Hall A left-HRS spectrometer with its septum magnet will detect the leading hadrons at $6°$ ($\Delta\Omega \approx 5$ msr) with a momentum of 2.40 GeV/c ($z_h = E_h/\nu \sim 0.5$) for either positive or negative polarity. The recently constructed BigBite spectrometer ($\Delta\Omega \approx 60$ msr) will be used as the electron detector at $30°$ to detect the scattered electrons with $0.8 \sim 2.1$ GeV/c in coincidence ($0.12 < x < 0.41$, $Q^2 = 1.21 \sim 3.14$ GeV^2). Since the π^- and π^+ phase spaces are identical and the detection efficiencies can be well-controlled, relative π^-/π^+ yield ratios can be easily determined such that the flavor non-singlet combination $A_{1n}^{\pi^+ - \pi^-}$ can be constructed.

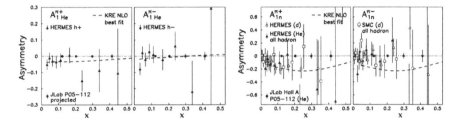

Figure 1. The expected statistical uncertainties of A_{1He}^π (left) and A_{1n}^π (right) of Jefferson Lab proposal P05-112. The SMC and the HERMES deuteron data [1,2] have been translated into neutron asymmetries assuming leading order x-z factorization. The dashed curves are from the next-to-leading order global fit [4] of the existing data.

For 30 days of beam time, assuming 75% beam polarization and 42% target polarization, the statistical accuracy of A_{1He}^π can be improved by an order of magnitude compared to earlier HERMES data [5], as shown in Fig. 1. Significant improvements are expected on A_{1n}^π when compared to the deuteron data from SMC and HERMES. Following the Christova-Leader method of Eq. 1, Δd_v can be extracted from $A_{1n}^{\pi^+ - \pi^-}$, as shown together in Fig. 2 with the HERMES data [2] from the purity method. When combined with the upcoming proton data [6] of JLab experiment E04-113, sea flavor asymmetries can be extracted at leading order following Eq. 3. With a factor of five improvement on statistical accuracy compared to that of the HERMES $\Delta\bar{u} - \Delta\bar{d}$ results, this experiment might provide the first opportunity to discover a possible polarized sea asymmetry.

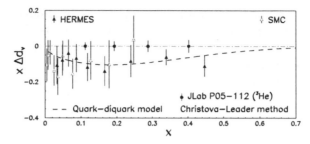

Figure 2. The statistical accuracy of Δd_v compared to the SMC [1] and the HERMES data [2]. The dashed curve is from a covariant quark-diquark model calculation [8] of Cloet et al.

Adding this set of ^{3}He data to the global NLO QCD fit, a factor of three improvement on sea quark polarization moments can be expected [4], as shown in Fig. 3. Indirectly, this data set will also improve the constraints on the gluon polarization Δg by a factor of three, comparable to the impact of the expected RHIC-2007 $A_{LL}^{\pi^0}$ data, as shown in Fig. 4. The reason for this sensitivity is because Δg is obtained in the global fit through the Q^2-evolutions of the inclusive structure functions g_1 which are coupled to the sea distribution. Once the valence distribution is reasonably separated from the sea with the SIDIS data, the gluon polarization can be better constrained in a global fit.

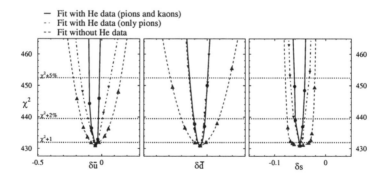

Figure 3. The expected improvement on the sea polarization moments in NLO global fit [4]. The dashed lines are the existing constraints, the dot-dashed and the solid lines are the constraints after adding pion data and/or kaon data from this proposal. The horizontal lines correspond to a deviation of $\chi^2 + 1$, $\chi^2(1 + 2\%)$ and $\chi^2(1 + 5\%)$ from the best fit.

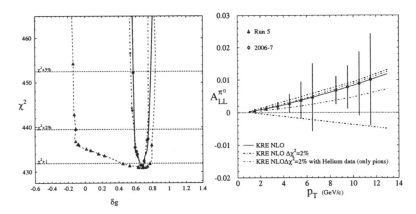

Figure 4. The constraint on the moment of the gluon polarization [4] by this measurement (left, curves as labled in Fig. 3) is compared with that from the expected RHIC-2007 $A_{LL}^{\pi^0}$ data (right). In the right panel, the area covered between the two inner blue dot-dashed lines corresponds to the $\chi^2(1 + 2\%)$ crossover region with the blue-dashed line on the left panel.

3. Conclusions

A Jefferson Lab Hall A experiment proposal to measure polarized ^{3}He SIDIS asymmetries in $\vec{n}(\vec{e}, e'h)X$ reactions was discussed. The proposed measurement will dramatically improve the precision of the world data set of A_{1n}^h and our knowledge of $\Delta d_v(x)$. Strong constraints on quark and gluon polarizations can be imposed through a global NLO QCD analysis.

Acknowledgments

The author thanks G. A. Navarro, R. Sassot, E. Christova, E. Leader, C. Weiss, J.-P. Chen, R. Gilman and J.-C. Peng for many discussions.

References

1. The Spin Muon Collaboration, *Phys. Lett.* B **420**, 180 (1998).
2. The HERMES collaboration, *Phys. Rev.* D **71**, 012003 (2005).
3. E. Christova and E. Leader, *Nucl. Phys.* B **607**, 369 (2001).
4. D. de Florian, G. A. Navarro and R. Sassot, *Phys. Rev.* D **71**, 094018 (2005).
5. The HERMES collaboration, *Phys. Lett.* B **464**, 123 (1999).
6. Jefferson Lab experiment E04-113, X. Jiang, P. Bosted, D. Day and M. Jones co-spokespersons, hep-ex/0412010.
7. Jefferson Lab experiment proposal P05-112, X. Jiang spokesperson.
8. I.C. Cloet, W. Bentz and A.W. Thomas, *Phys. Lett.* B **621**, 246 (2005).

SPIN STRUCTURE FUNCTIONS: PROTON/DEUTERON MEASUREMENTS IN THE RESONANCE REGION

M. K. JONES FOR THE *RSS* COLLABORATION*

Jefferson Lab
12000 Jefferson Ave.
Newport News, VA 23606, USA
E-mail: jones@jlab.org

The *RSS* experiment ran in Hall C at Jefferson Lab and measured the proton and deuteron beam-target asymmetries for parallel and perpendicular target fields over a W range from pion threshold to 1.9 GeV at $Q^2 \approx 1.3$ GeV^2. Preliminary results for the proton spin structure functions g_1 and g_2 are presented.

1. Introduction

This workshop has highlighted the large amount of data that is available for studying quark-hadron duality in the unpolarized structure functions and making precision tests of duality (see review [1]). With precision measurements in the resonance region, examination of duality in the g_1 and g_2 spin structure functions (SSFs) can also be done. A model independent determination of g_1 and g_2 can be done with measurement of the beam-target asymmetry at two different orientations of the target spin. g_1 has been measured in the DIS region for a wide range of Q^2. At SLAC, g_2 and g_1 were measured in the resonance region, but with large W bins. In Hall B at Jefferson Lab, precision data for the parallel beam-target asymmetries on the proton and deuteron in the DIS and resonance regions have been measured[2,3,4] over a large range of moderate Q^2 and g_1 was extracted using estimates of g_2. Recent measurements[5] of neutron g_1 and g_2 in the resonance region using a polarized ^{3}He target have been done at Jefferson Lab

---

*RSS collaboration (Jlab E01-006): U. Basel, Florida International U., Hampton U., U. of Massachusetts, U. of Maryland, Mississippi State U., North Carolina A&T STATE U., U. of N. C. at Wilmington, Norfolk State U., Old Dominion U., S.U. at New Orleans, U. of Tel-Aviv, Jefferson Lab, U. of Virginia, Virginia P. I. & S.U., Yerevan Physics Institute. Spokesmen: O. A. Rondon and M. K. Jones .

at $Q^2 < 1$ and preliminary results of another Hall A experiment[6], which studied the dependence of neutron g_1 and g_2 in the resonance region for $1 < Q^2 < 4$ GeV^2, have been presented at this workshop. A lack of data for the proton g_2 in the resonance region at low Q^2 is a glaring hole in the kinematic map of SSFs. The goal of the Hall C *RSS* experiment is to measure the W dependence of proton and deuteron g_1 and g_2 in the resonance region at a $Q^2 \approx 1.3$ GeV^2.

In the *RSS* experiment, beam-target asymmetries were measured for proton and deuteron targets with the target spin parallel ($A_{\parallel}$) and perpendicular ($A_{\perp}$) to the beam helicity. The $A_{\parallel}$ and $A_{\perp}$ can be combined to determine the spin structure functions g_1 and g_2 or the related spin asymmetries A_1 and A_2. Duality can be studied for both g_1 and g_2 and, with precision data in small W bins, local quark-hadron duality can be investigated. While g_2 does not have a simple partonic interpretation, g_2 is sensitive to higher twist.

2. Experiment

The *RSS* experiment was performed in Hall C at the Thomas Jefferson National Accelerator Facility (Jefferson Lab). Polarized electrons with 5.76 GeV/c momentum were scattered from polarized frozen ammonia (NH$_3$) and deuterated ammonia (ND$_3$) targets. The spin of the polarized target was aligned anti-parallel and perpendicular to the beam. The scattered electrons were detected at 13.15° in the High Momentum Spectrometer (HMS). Electron particle identification was done by a combination of a gas Cerenkov detector and lead-glass calorimeter. To cover the range of W from 0.6 GeV to 1.9 GeV, data was taken with the HMS at two central momentum settings of 4.73 and 4.09 GeV/c. The average momentum transfer-squared, Q^2, was centered at 1.3 ± 0.3 GeV^2 for $1.1 < W < 1.9$. GeV.

The ammonia targets were polarized by dynamic nuclear polarization and operated at 1 K in a 5 T magnetic field. To maintain reasonable target polarization, the beam current was limited to <100 nA and was uniformly rastered over a 2 cm diameter circle. The target polarization , P_T, was measured continuously by the NMR technique. The average proton polarization was 70% and the average deuteron polarization was 20%. The polarization of the beam, P_B, was measured in Hall C using the Moller polarimeter. The average P_B was 65.6% (71%) for perpendicular (parallel) target field.

The measured asymmetry, A_m, is defined as $\frac{N^+-N^-}{N^++N^-}$ where N^+ and N^- are the raw counts normalized for deadtime and charge for opposite beam helicities. The asymmetry for parallel and perpendicular target field is

$$A_{\parallel,\perp} = \frac{1}{C_N f_{RC}}(\frac{A_m}{f P_B P_T} - C_D) + A_{RC} \tag{1}$$

where the measured asymmetry is normalized by P_T, P_B and the dilution factor, f. The dilution factor is the ratio of the yield from polarized free protons to yield from all target material. The asymmetry is also corrected for radiative effects (f_{RC} and A_{RC}) and the contribution from the small nitrogen polarization (C_N, C_D). The C_N correction is ≈ 1.02 has not been applied to the preliminary data that is being shown. The C_D correction is for deuteron target only.

Elastic ep asymmetry, A_{ep}, was measured when the HMS central momentum was 4.73 GeV/c. The sensitivity of A_{ep} to the ratio of the proton electric to magnetic form factor, $\frac{G_E}{G_M}$, depends on angle between the target spin and the momentum transfer. For parallel target field, A_{ep} is insensitive to $\frac{G_E}{G_M}$ and was used to determine $P_B P_T$ to a combined relative statistical and systematic error of 1.9%. For the perpendicular target field, the sensitivity of A_{ep} to $\frac{G_E}{G_M}$ is large enough that $\frac{G_E}{G_M}$ can be determined from A_{ep} using $P_B P_T$ determined by NMR technique and Moller polarimeter.

3. Results

Preliminary results for the proton spin structure functions, g_1^p and g_2^p, from the RSS experiment are plotted versus Bjorken x_{bj} in Fig. 1a and Fig. 1b respectively. In Fig. 1a, previous g_1^p data[7] from SLAC E143 at $Q^2 \approx 5$ GeV^2 and data[2] from Jefferson Lab Hall B at $Q^2 \approx 1.35$ GeV^2 are plotted for comparison. While agreement is good for $x_{bj} > 0.5$, the RSS data exhibit more structure in the $0.5 > x_{bj} > 0.3$ region compared to the Hall B data. But at this time, the RSS data are still preliminary and radiative corrections are being finalized. The RSS g_1^p are compared with calculations from three different PDFs (BSB[9], GRSV[10] and AAC[11]) which have been evolved to $Q^2 = 1.3$ GeV^2. The GRSV and AAC PDFs have target mass corrections. The RSS data oscillate around the PDF curves.

In Fig. 1b, arrows indicate the associated W for the given x_{bj} for the RSS experiment. Here, data[8] for g_2^p from SLAC E155 are plotted. The SLAC data are at higher Q^2 and in the DIS region. The RSS g_2^p has significant structure in the $0.5 > x_{bj} > 0.3$ region.

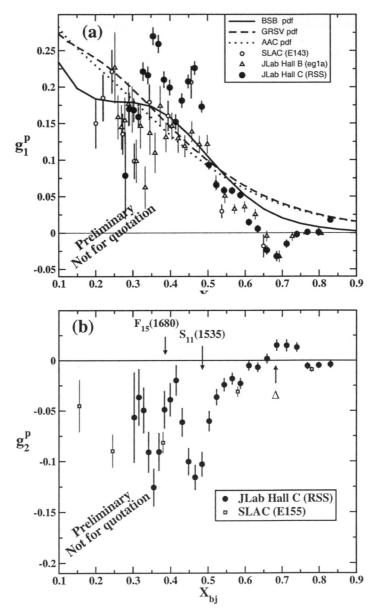

Figure 1.   The proton structure functions g_1^p and g_2^p. The curves are described in the text.

Once the data are finalized, ratios of data to the PDFs can be made that will enable quantitative tests of duality in proton SSFs. In addition, the *RSS* experiment measured deuteron beam-target asymmetries and these data will be combined with the proton data to extract neutron SSFs.

4. Acknowledgments

This work was supported in part by the U. S. Department of Energy (DOE) contract DE-AC05-84ER40150, under which the Southeastern Universities Research Association (SURA) operates the Thomas Jefferson National Accelerator Facility (Jefferson Lab).

References

1. W. Melnitchouk, R. Ent and C. Keppel, *Phys. Rept.* **406**, 127 (2005).
2. R. Fatemi *et al.*, *Phys. Rev. Lett.* **91**, 222002 (2003).
3. J. Yun *et al.*, *Phys. Rev.* **C67**, 055204 (2003).
4. Y. Prok, *These proceedings.*
5. M. Amarian *et al.*, *Phys. Rev. Lett.* **92**, 022301 (2004).
6. P. Solvignon, *These proceedings.*
7. K. Abe *et al.*, *Phys. Rev.* **D58**, 112003 (1998).
8. P. Anthony *et al.*, *Phys. Lett.* **B533**, 17 (2003).
9. C. Bourrely, J. Soffer and F. Buccella, *Eur. Phys. J.* **C23**, 487 (2002).
10. M. Glueck, E. Reya, M. Stratmann and W. Vogelsang, *Phys. Rev.* **D63**, 094005 (2001).
11. M. Hirai *et al.*, *Phys. Rev.* **D69**, 054021 (2004).

Duality in Photoproduction

Duality in Perturbation

DUALITY IN VECTOR-MESON PRODUCTION

A. DONNACHIE

School of Physics and Astronomy,
University of Manchester,
Manchester M13 9PL, England

The origins of duality in pion-nucleon scattering are recalled. A simple model of vector-meson photoproduction is outlined based on vector-meson dominance. This gives an excellent description of data at small t and, analogously to pion-nucleon scattering, it can be concluded that vector-meson photoproduction also satisfies the original duality. The situation for vector-meson electroproduction is more complicated and it is argued that, in the kinematical domain acessible to HERMES and JLab, nonperturbative effects may be significant. This has implications for the study of generalized parton distributions in vector-meson electroproduction.

1. Introduction

It is convenient to start by recalling the origins of duality in pion-nucleon scattering through finite-energy sum rules (FESRs). These relate an integral over the resonance region at fixed t to a sum over the Regge-pole terms appropriate to higher energies. As the FESRs are written for amplitudes, knowledge of the low-energy amplitudes, from partial-wave analysis, provides important information about the Regge-pole amplitudes. It is necessary to assume that the Regge-pole amplitude describes the real physical amplitude at low energy on the average and this does happen in practice. The original example [1,2] compared $p_{\mathrm{Lab}}(\sigma^{\mathrm{Tot}}(\pi^- p) - \sigma^{\mathrm{Tot}}(\pi^+ p))$ with the Regge fit to high-energy data. In this case it is the exchange of the ρ trajectory, extrapolated to low energies.

The $\pi^- p$ and $\pi^+ p$ high-energy elastic scattering amplitudes receive equal contributions from pomeron exchange, which cancels in the difference between the two total cross sections. This implies that the non-pomeron Regge-pole t-channel exchanges are dual to the s-channel resonances. The extrapolations to low energy of the Regge fits to the high-energy $\pi^+ p$ and $\pi^- p$ total cross sections give good descriptions of the average low-energy cross sections in each case [2]. The resonances sit on a non-resonance background, so assuming that the non-pomeron t-channel exchanges are dual to the low-energy s-channel resonances leads to the assumption that pomeron exchange is dual to the low-energy s-channel non-resonance background.

This asumption of two-component duality is explicitly realized in the individual partial-wave amplitudes in πN scattering [4]. The linear combinations of s-channel partial-wave amplitudes corresponding to isospin 1 exchange in the t-channel are given entirely by s-channel resonances. For the linear combinations of s-channel partial-wave amplitudes corresponding to isospin 0 exchange in the t-channel, the s-channel resonances are superimposed on a predominantly-imaginary smooth background.

This is the orginal form of duality and was used in the analysis of pion-nucleon scattering [3,4] and pion photoproduction [5,6,7,8]. Using simple vector-meson dominance (VMD) arguments it can be shown that vector-meson photoproduction also satisfies this original form of duality. We then explore the question of extended duality for vector-meson electroproduction, and the relative roles of perturbative and nonperturbative QCD.

2. Vector-meson photoproduction

A direct connection between πp scattering and ρ^0 and ω photoproduction is provided via the assumption of VMD [9]. In its simplest form, VMD says that the cross section for $\gamma p \to \rho^0 p$ is given by

$$\frac{d\sigma}{dt}(\gamma p \to \rho^0 p) = \alpha \frac{4\pi}{\gamma_\rho^2} \frac{d\sigma}{dt}(\rho^0 p \to \rho^0 p) \tag{1}$$

where $4\pi/\gamma_\rho^2$ is the ρ-photon coupling, which can be found from the e^+e^- width of the ρ^0. The amplitude for $\rho^0 p \to \rho^0 p$ is simply given by the average of the amplitudes for $\pi^- p$ and $\pi^+ p$ elastic scattering. In this combination of πp scattering amplitudes the $C = -1$ exchanges cancel (as they should) leaving the pomeron and f_2 exchanges. There are two omissions in this procedure. One is a contribution from a_2 exchange, which is forbidden in πp elastic scattering by g-parity but is allowed for ρ^0 photoproduction. However this is expected to be extremely small as the a_2-nucleon coupling is much weaker than the f_2-nucleon coupling and as the photon is necessarily isoscalar (ω-like) while for f_2 exchange the photon is isovector (ρ-like). The net effect is that the amplitude for a_2 exchange is a few percent of that for f_2 exchange. The other omission is pion exchange. This is again small as the photon is once more necessarily isoscalar and its contribution decreases rapidly with increasing energy, so it is relevant only close to threshold.

The trajectories of the pomeron, f_2 and a_2 are well-known from hadronic scattering, as is the mass scale by which we must divide s before raising it to the Regge power. It is well-established [2] that the trajectories couple to the proton through the Dirac electric form factor $F_1(t)$. Wherever it can be experimentally checked, the differential cross section for ρ^0 photoproduction is found to have the same slope at small t as the $\pi^{\pm}p$ elastic differential cross sections, so it is natural to assume that the form factor of the ρ, $F_\rho(t)$, is the same as that of the pion.

The amplitude for $\gamma p \to \rho^0 p$ is then uniquely defined using (1) and is given by [10]

$$T(s,t) = iF_1(t)F_\rho(t)\Big(A_P(\alpha'_P s)^{\alpha_P(t)-1}e^{-\frac{1}{2}i\pi(\alpha_P(t)-1)}$$

$$+ A_R(\alpha'_R s)^{\alpha_R(t)-1}e^{-\frac{1}{2}i\pi(\alpha_R(t)-1)}\Big) \qquad (2)$$

with the coefficients A_P and A_R obtained [10] from the $\rho^0 \to e^+e^-$ width and a standard fit to the $\pi^{\pm}p$ total cross sections. The resulting predictions [10] for the $\gamma p \to \rho^0 p$ differential cross section agree with data at small t from $\sqrt{s} = 4.3$ GeV [11] to $\sqrt{s} = 71.7$ and 94 GeV [12].

This model has implications for polarization effects in ρ^0 photoproduction. It is well known in ρ^0 photoproduction that the helicity of the ρ^0 is the same as that of the photon, the phenomenon of s-channel helicity conservation (SCHC). What is less well known is that pomeron and f_2 exchange also conserve helicity equally well at the nucleon vertex, a result deduced [2] from polarized-target asymmetries in $\pi^{\pm}p$ scattering. Target polarization effects in ρ^0 photoproduction will arise primarily from the interference of the dominant pomeron and f_2 exchange with the a_2 and other small unknown exchanges. This does not necessarily mean that polarized-target asymmetries will be small: they are rather large at low and intermediate energies in $\pi^{\pm}p$ elastic scattering through interference of the dominant pomeron and f_2 exchange with ρ exchange, which is spin flip. It does mean that they are not predictable and that they essentially measure the small amplitudes.

The discussion on ρ^0 photoproduction can be applied directly to ω photoproduction with three numerical modifications. The cross section is approximately a factor of 9 smaller due to the difference between $4\pi/\gamma^2_\rho$ and $4\pi/\gamma^2_\omega$, the a_2 contribution is larger as the photon is now ρ-like for a_2 exchange and the cross section from pion exchange is larger by a factor of 9 than in ρ^0 photoproduction. By using plane-polarized photons [11] the natural-parity ($J^P = (-1)^J$) and unnatural parity ($J^P = (-1)^{J+1}$) exchanges can be separated. This confirms that pion-exchange is the dominant contribution near threshold and shows that the cross-section from $C = -1$ exchange is is larger than the cross-section from $C = +1$ exchange until $\sqrt{s} \approx 2.5$ GeV. At energies above $\sqrt{s} \approx 4.5$ GeV the cross section is well described in magnitude and shape by (1) with the replacement of $4\pi/\gamma_\rho^2$ by $4\pi/\gamma_\omega^2$. So apart from near threshold, ρ^0 and ω photoproduction are described by the same physics and the process is hadron-like at small t.

For ϕ photoproduction, because of the pomeron dominance arising from Zweig's rule, the cross section should behave as $s^{2\epsilon}/b$ where b is the near-forward t-slope. The data are compatible with this, but are not sensitive to constant b or to letting the forward peak shrink in the canonical way, i.e. by taking $b = b_0 + 2\alpha' ln(\alpha' s)$.

The model predictions for ρ^0 photoproduction [2,10] show good agreement with data for $|t| \leq 1.0$ GeV^2 (the experimental limit) at low and intermediate energies but at $\sqrt{s} = 71.7$ and 94 GeV the agreement is good only out to $|t| = 0.5$ GeV^2. At larger values of $|t|$ the predicted cross section falls below the data, the discrepancy increasing with increasing $|t|$. To describe the proton structure function $F_2(x, Q^2)$ at small x within the framework of conventional Regge theory it is necessary to introduce [13] a second pomeron, the hard pomeron, with intercept a little greater than 1.4. This concept is also compatible [14] with the ZEUS data [15] for the charm component $F_2^c(x, Q^2)$ of $F_2(x, Q^2)$ which seem to confirm the existence of the hard pomeron. The slope of the trajectory can be deduced [2,14] from the H1 data [16] for the differential cross section for the process $\gamma p \to J/\psi p$. Fits to the data for F_2 at small x and for F_2^c suggest that the coupling of the hard pomeron to quarks is flavour-blind, so the hard-pomeron contribution to $\gamma p \to \rho^0 p$ can be obtained from that in $\gamma p \to J/\psi p$ by including the effect of the vector-meson wave functions. The result of adding this hard-pomeron contribution to the amplitude for $\gamma p \to \rho^0 p$ at $\sqrt{s} = 71.7$ and 94 GeV gives an excellent description of the data. The hard pomeron is also necessary for and compatible with the data on $\gamma p \to \phi p$ at $\sqrt{s} = 71.7$ and 94 GeV. Of course at low and intermediate energies this hard-pomeron term will be negligible compared to the soft pomeron because of its very strong energy dependence.

3. Vector-meson electroproduction

In the laboratory frame, the incoming photon develops hadronic fluctuations some distance from the proton target. The typical distance travelled by a vacuum fluctuation of a photon into a hadron is large compared to the size of the proton or the range of the strong interaction, even for a photon laboratory energy of 10 GeV. Thus the amplitude for vector-meson photo-production can be factorized into an amplitude for the conversion of the photon to the vector meson times the amplitude describing the interaction of the vector meson with the proton target. This is the basis of the VMD approach described in the previous section. The same space-time picture can be applied at large Q^2, provided that s is sufficiently large. However at large Q^2 it is necessary to include many vector mesons in the photon fluctuation and the simplicity of (1) is lost. This is analogous to the situation in e^+e^- annihilation. At energies below 2.5 GeV or so, e^+e^- annihilation can be described by a limited number of vector mesons, but at higher energies it is much more convenient (and sensible) to consider the reaction as $e^+e^- \to q\bar{q}$. So in vector-meson electroproduction it is more sensible to consider the photon fluctuating into a $q\bar{q}$ pair which then scatters on the target proton and then recombines into the vector meson (or indeed any other hadronic final state). This is the basis of the dipole model and its extension to generalized parton distributions.

This can be directly observed in the variation of the forward slope, b, of $d\sigma/dt$ for $\gamma^*p \to \rho^0 p$ as a function of Q^2. For example, at $\sqrt{s} = 10$ GeV, it falls from the typical hadronic value of 9–10 GeV^{-2} at $Q^2 = 0$ to about 5 GeV^{-2} at $Q^2 = 15$ GeV^2. This latter value for b corresponds rather closely to what one would expect from (2) without $F_\rho(t)$. In other words there is no contribution to the differential cross section from structure at the photon-ρ^0 transition vertex. Thus the reaction goes from being purely nonperturbative at $Q^2 = 0$ to being purely perturbative for $Q^2 \gtrsim 15$ GeV^2.

This apparent "mix" of hadron-like behaviour and perturbative behaviour at moderate Q^2 is easily seen from the energy dependence of the cross section. A good example is provided by the data at $\langle Q^2 \rangle = 3.5$ GeV^2, which clearly show the rapid decrease with increasing energy due to Regge exchange followed by the increase due to pomeron exchange. As Q^2 increases the energy dependence at high energy increases [17]. If the data are parametrized as s^δ, then δ increases from a value compatible with soft-pomeron exchange for $Q^2 \approx 0$ to a value consistent with hard-pomeron exchange for $Q^2 \gtrsim 20$ GeV^2.

Is it correct to make this separation between perturbative and nonperturbative QCD or does duality in the Bloom-Gilman sense, as applied to the proton structure function, render this unnecessary? This is the key question, as yet unanswered, for the use of vector-meson electroproduction in providing information on GPDs. There is a significant difference between determining F_2 and extracting a GPD. The former is obtained directly from the cross section, the latter requires model assumptions. In the kinematical regime accessible to HERMES and to the JLab upgrade, the standard GPD perturbative amplitudes may not be the whole story for vector meson electroproduction. Is the hadron-like photon dual to higher twist and/or to power corrections? A similar question has been raised [18] in the context of $\gamma^*\gamma \to \rho^0\rho^0$

In two-gluon exchange models, the energy dependence of ρ^0 electroproduction at high energy reflects the proton's gluon distribution. In dipole models it appears through the structure of the dipole cross section or by explicitly introducing soft and hard pomeron terms. The limit of $Q^2 \to 0$ is handled by modifying the photon wave function, for example through a Q^2-dependent quark mass, which simulates the hadron-like nature of the photon at small Q^2 but introduces additional model dependence. Note that there is ambiguity between the choice of wave functions and the details of the reaction mechanism and represents an additional problem for the etxraction of GPDs. The sensitivity to the choice of wave function persists [19] for the J/ψ and even for the Υ.

References

1. K. Igi and S. Matsuda, *Phys. Rev. Lett.* **18**, 625 (1967)
2. A. Donnachie, H. G. Dosch, P. V. Landshoff and O. Nachtmann, *Pomeron Physics and QCD*, Cambridge University Press 2002.
3. V. Barger and R. J. N. Phillips, *Phys. Rev.* **187**, 2210 (1969).
4. H. Harari and Y. Zarmi, *Phys. Rev.* **187**, 2230 (1969).
5. R. Worden, *Nucl. Phys.* **B37**, 253 (1972).
6. I. S. Barker, A. Donnachie and J. K. Storrow, *Nucl. Phys.* **B95**, 347 (1975).
7. I. S. Barker and J. K. Storrow, *Nucl. Phys.* **B137**, 413 (1978).
8. M. Rahnama and J. K. Storrow, *J. Phys.* **G17**, 243 (1991).
9. A. Donnachie and G. Shaw in *Electromagnetic Interactions of Hadrons* vol 2, A Donnachie and G Shaw eds, Plenum Press (1978).
10. A. Donnachie and P. V. Landshoff, *Phys. Lett.* **B348**, 213 (1995)
11. J. Ballam *et al*, *Phys. Rev.* **D7**, 3150 (1973).
12. J. Breitweg *et al* (ZEUS Collaboration), *Eur. Phys. J.* **C1**, 81 (1998).
13. A. Donnachie and P. V. Landshoff, *Phys. Lett.* **B437**, 408 (1998).
14. A. Donnachie and P. V. Landshoff, *Phys. Lett.* **B478**, 146 (2000).
15. J. Breitweg *et al* (ZEUS Collaboration), *Eur. Phys. J.* **C14**, 213 (2000).
16. C. Adloff *al* (H1 Collaboration), *Phys. Lett.* **B483**, 23 (2000).
17. J. Breitweg *et al* (ZEUS Collaboration), *Eur. Phys. J.* **C6**, 603 (1999).
18. I. V. Anikin, B. Pire and O. V. Teryaev, *Phys. Rev.* **D69**, 014018 (2004)
19. H. G. Dosch and E. Ferreira, in preparation

PHOTOPION PRODUCTION FROM NUCLEON AND SCALING

H. GAO, D. DUTTA

Triangle Universities Nuclear Laboratory and Department of Physics,
Duke University,
Durham, NC 27708, USA
E-mail: gao@phy.duke.edu
ddutta@tunl.duke.edu

In this talk, we present the most recent data on charged pion photoproduction
from nucleon at Jefferson Lab. These new data suggest a transverse momentum of
~ 1.2 (GeV/c) as the physical quantity governing the onset of the scaling behavior.
We also present a new analysis of the proton-proton elastic scattering data and
also of the photopion production data based on the generalized counting rule.

The transition between perturbative and non-perturbative regimes of
Quantum Chromo Dynamics (QCD) is of long-standing interest in nuclear
and particle physics. Exclusive processes play a central role in studies trying
to map out this transition. The differential cross sections for many exclusive
reactions [1] at high energies and large momentum transfers appear to obey
dimensional scaling laws [2] (also called quark counting rules). In recent
years, the onset of this scaling behavior has been observed at a hadron
transverse momentum of ~ 1.2 (GeV/c) in deuteron photo-disintegration
[3,4]. Measurements on photopion production from nucleon have been carried
out recently at Jefferson Lab [5,6] as shown in Figure 1. These results agree
with the world data within uncertainties in the overlapping region. The
data at $\theta_{cm} = 70°, 90°$ exhibit a global scaling behavior predicted by the
constituent counting rule in π^- channel, similar to what was observed in
the π^+ channel at similar center-of-mass angles. The data at $\theta_{cm} = 50°$ do
not display scaling behavior and may require higher photon energies for the
observation of the onset of the scaling behavior. These data suggest that
a transverse momentum of around 1.2 GeV/c might be the scale governing
the onset of scaling for the photo-pion production, which is consistent with
what has been observed in deuteron photodisintegration [3,4]. Data in these

90

two channels at 90° show possible oscillations [5] around the scaling behavior in similar ways.

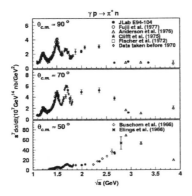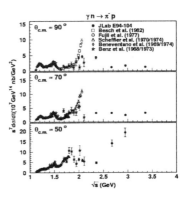

Figure 1. *The scaled differential cross section* $s^7 \frac{d\sigma}{dt}$ *versus center-of-mass energy for the* $\gamma p \to \pi^+ n$ *(upper panel) and* $\gamma n \to \pi^- p$ *(lower panel) at* $\theta_{cm} = 90°, 70°, 50°$.

On the other hand, those models which predict scaling law also predict hadron helicity conservation in exclusive processes [7]. However, experimental data in similar energy and momentum regions tend not to agree with these helicity conservation selection rules [8]. Although contributions from non-zero parton orbital angular momenta are power suppressed, as shown by Lepage and Brodsky [9], they could break hadron helicity conservation rule [10]. Interestingly, recent re-analysis of quark orbital angular momenta seems to contradict the notion of power suppression [11]. Furthermore, Isgur and Llewellyn Smith [12] argued that non-perturbative processes could still be important in some kinematic regions even at high energies. Thus the transition between the perturbative and non-perturbative regimes remains obscure and makes it essential to understand the exact mechanism governing the early onset of scaling behavior.

Towards this goal, it is important to look closely at claims of agreement between the differential cross section data and the quark counting rule prediction. Deviations from the quark counting rules have been found in exclusive reactions such as elastic proton-proton (pp) scattering [13,14]. In fact, the re-scaled 90° center-of-mass pp elastic scattering data ($s^{10} \frac{d\sigma}{dt}$), show substantial oscillations about the power law behavior. Oscillations are not restricted to the pp elastic scattering channel; they are also seen in elastic πp fixed angle scattering [15] and hints of oscillation about the s^{-7} scaling have also been reported in the recent data [5] from Jefferson Lab (JLab) on photo-pion production above the resonance region. In addition to violations of the scaling laws, spin correlations in polarized pp elastic scattering also show significant deviations from perturbative QCD (pQCD) expectations [16,17]. Several sets of arguments have been put forward to account for these deviations from scaling laws and the unexpected spin correlations. Brodsky and de Teramond [18] explain the pp scattering data in terms of the opening up of the charm channel and excitation of $c\bar{c}uuduud$ resonant states. Alternatively the deviations are said to be an outcome of the interference between the pQCD (short distance) and the long distance Landshoff amplitude (arising from multiple independent scattering between quark pairs in different hadrons) [19]. Gluonic radiative corrections to the Landshoff amplitude give rise to an energy dependent phase [20] and thus the energy dependent oscillation. Carlson, Chachkhunashvili, and Myhrer [21] have also applied a similar interference concept to explain the pp polarization data. The QCD re-scattering calculation of the deuteron photo-disintegration process by Frankfurt, Miller, Sargsian and Strikman [22] predicts that the additional energy dependence of the differential cross-section, beyond the $\frac{d\sigma}{dt} \propto s^{-11}$ scaling, arises primarily from the $n - p$ scattering in the final state. In this scenario the oscillations may arise due to QCD final state interaction. If these predictions are correct, such oscillatory behavior may be a general feature of high energy exclusive photo-reactions.

Recently, a number of new developments have generated renewed interest in this topic. Zhao and Close [23] have argued that a breakdown in the locality of quark-hadron duality (dubbed as "restricted locality" of quark-hadron duality) results in oscillations around the scaling curves predicted by the counting rule. They explain that the smooth behavior of the scaling laws arise due to destructive interference between various intermediate resonance states in exclusive processes at high energies. However, at lower energies this cancellation due to destructive interference breaks down locally and gives rise to oscillations about the smooth behavior. On the other hand, Ji et al. [24] have derived a generalized counting rule based on a pQCD inspired model, by systematically enumerating the Fock components of a hadronic light-cone wave function. Their generalized counting rule for hard exclusive processes include parton orbital angular momentum

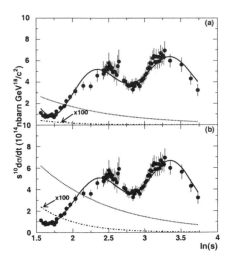

Figure 2. (a) The fit to pp scattering data at $\theta_{cm} = 90°$ when helicity flip amplitudes are included as described in [30] . The parameters for the energy dependent phase was kept same as the earlier fit of Ralston and Pire [19]. The solid line is the fit result, the dotted line is contribution from the helicity flip term $\sim s^{-11}$, the dot-dashed line is contribution from the helicity flip term $\sim s^{-12}$. The $\sim s^{-12}$ contribution has been multiplied by 100 for display purposes.(b) The same data fitted to the form described in [30] but with the new more general parametrization of the Landshoff amplitude.

and hadron helicity flip, thus they provide the scaling behavior of the helicity flipping amplitudes. The interference between the different helicity flip and non-flip amplitudes offers a new mechanism to explain the oscillations in the scaling cross-sections and spin correlations. The counting rule for hard exclusive processes has also been shown to arise from the correspondence between the anti-de Sitter space and the conformal field theory [25] which connects superstring theory to conformal gauge theory. Brodsky *et al.* [26] have used this anti-de Sitter/Conformal Field Theory correspondence or string/gauge duality to compute the hadronic light front wave functions. This yields an equivalent generalized counting rule without the use of perturbative theory. Moreover, pQCD calculations of the nucleon form factors including quark orbital angular momentum [28,27] and those computed from light-front hadron dynamics [26] both seem to explain the $\frac{1}{Q^2}$ fall-off of the proton form-factor ratio, $G_E^p(Q^2)/G_M^p(Q^2)$, measured recently at JLab in polarization transfer experiments [29].

Recently, we have shown [30] that the generalized counting rule of Ji *et al.* [24] along with the Landshoff terms and associated interferences does a better job of describing the oscillations about the quark counting rule, in the pp elastic scattering data at $\theta_{cm} = 90°$. Fig 2 shows the results of our fit and also shows the explicit contributions from the s^{-11} and s^{-12} term for this approach. The value of Λ_{QCD} was fixed at 100 MeV for all fits. This new fit is in much better agreement with the data. This is specially true in the low energy region ($s < 10$ GeV^2). The helicity flip amplitudes (mostly the term $\sim s^{-4.5}$) are significant at low energies and seem to help in describing the data at low energies. The contributions from helicity flipping amplitudes which are related to quark orbital angular momentum, seem to play an important role at these low energies, which is reasonable given that the quark orbital angular momentum is non-negligible compared to the momentum scale of the scattering process. It is interesting to note that among the helicity flip amplitudes the one with the lower angular momentum dominates. Similarly the spin-correlation A_{NN} in polarized pp elastic scattering data can be better described [30] by including the helicity flipping amplitude along with the Landshoff amplitude and their interference. The photo-pion production data from nucleons at large angles can also be described similarly [30]; however, because of the coarse energy spacing of the data, the results are not as illustrative. This points to the urgent need for more data on pion-photoproduction above the resonance region with finer energy spacing.

We thanks Dr. L.Y. Zhu for providing Figure 1. This work is supported by the U.S. Department of Energy under contract number DE-FC02-94ER40818 and DE-FG02-03ER41231.

References

1. C. White *et al.*, Phys. Rev. **D49**, 58 (1994).
2. S. J. Brodsky and G.R. Farrar, Phys. Rev. Lett.**31**, 1153 (1973); Phys. Rev. D **11**, 1309 (1975); V. Matveev *et al.*, Nuovo Cimento Lett. **7**, 719 (1973);
3. C. Bochna *et al.*, Phys. Rev. Lett. **81**, 4576 (1998); E.C. Schulte, *et al.*, Phys. Rev. Lett. **87**, 102302 (2001);
4. P. Rossi *et al.*, Phys. Rev. Lett **94**, 012301 (2005); M. Mirazita *et al.*, Phys. Rev. C **70**, 014005 (2004).
5. L. Y. Zhu *et al.*, Phys. Rev. Lett. **91**, 022003 (2003)
6. L. Y. Zhu *et al.*, Phys. Rev. C **71**,044603 (2005); nucl-ex/0409018.
7. S. J. Brodsky and G. P. Lepage, Phys. Rev. D **24**, 2848 (1981).
8. K. Wijesooriya, *et al.*, Phys. Rev. Lett. **86**, , (2)975 (2001).

9. G. P. Lepage, and S. J. Brodsky, Phys. Rev. D **22**, 2157 (1980).

10. T. Gousset, B. Pire and J. P. Ralston, Phys. Rev. D **53**, 1202 (1996).

11. J. P. Ralston and P. Jain, Phys. Rev. D **69**, 053008 (2004).

12. N. Isgur and C. H. Llewellyn Smith, Phys. Rev. Lett. **52**, 1080 (1984).

13. C. W. Akerlof, *et al.*, Phys. Rev. 159, 1138 (1967); R. C. Kammerud, *et al.*, Phys Rev. D **4**, 1309 (1971); K. A. Jenkins, *et al.*, Phys. Rev. Lett, **40**, 425 (1978).

14. A.W. Hendry, Phys. Rev. D **10**, 2300 (1974).

15. D. P. Owen *et al.*, Phys. Rev. **181**, 1794 (1969); K. A. Jenkins *et al.*, Phys. Rev. D **21**, 2445 (1980); C. Haglin *et al.*, Nucl. Phys. B **216**, 1 (1983).

16. D.G. Crabb *et al.*, Phys. Rev. Lett. **41**, 1257 (1978).

17. G.R. Court *et al.*, Phys. Rev. Lett. **57**, 507 (1986);T.S. Bhatia *et al.*, Phys. Rev. Lett. **49**, 1135 (1982); E.A. Crosbie *et al.*, Phys. Rev. D **23**, 600 (1981).

18. S. J. Brodsky, and G. F. deTeramond; Phys. Rev. Lett. **60**, 1924 (1988).

19. J. P. Ralston and B. Pire, Phys Rev. Lett. **49**, 1605 (1982); B. Pire and J. P. Ralston, Phys. Lett. B **117**, 233 (1982)

20. A. Sen, Phys. Rev. D **28**, 860 (1983).

21. C.E. Carlson, M. Chachkhunashvili, and F. Myhrer, Phys. Rev. D **46**, 2891 (1992).

22. L.L. Frankfurt, G.A. Miller, M.M. Sargsian, and M.I. Strikman, Phys. Rev. Lett. **84**, 3045 (2000).

23. Q. Zhao and F. E. Close, Phys. Rev. Lett. **91**, 022004 (2003).

24. X. Ji, J.-P. Ma and F. Yuan, Phys. Rev. Lett. **90**, 241601 (2003).

25. J. Polchinski and M.J. Strassler, Phys. Rev. Lett. **88**, 031601 (2002); R.C. Brower and C.I. Tan, Nucl. Phys. B **662**, 393 (2003); O. Andreev, Phys. Rev. D **67**, 046001 (2003).

26. S. J. Brodsky and G. F. de Teramond, Phys. Lett. **B582**, 211 (2004); S. J. Brodsky,J. R. Hiller, D. S. Hwang and V. A. Karmanov, Phys. Rev. D **69**, 076001 (2004).

27. R. Buniy, J. P. Ralston and P. Jain in *VII International Conference on the Intersections of Particle and Nuclear Physics*, Quebec City, 2000, edited by Z. Parsa and W. Marciano (AIP, New York, 2000), hep/ph/0206074

28. A. V. Belitsky, X. Ji and F. Yuan, Phys. Rev. Lett. **91**, 092003 (2003).

29. M. K. Jones *et al.*, Phys. Rev. Lett. **84**, 1398 (2000); O. Gayou *et al.*, Phys. Rev. Lett. **88**, 092301 (2002).

30. D. Dutta and H. Gao, Phys. Rev. C **71**, 032201(R) (2005).

ONSET OF SCALING IN EXCLUSIVE PROCESSES

M. MIRAZITA

I.N.F.N. Laboratori Nazionali di Frascati,
Via E. Fermi 40 I-00044, Frascati, Italy
E-mail: mirazita@lnf.infn.it

The onset of pQCD description of deuteron reactions with electromagnetic probes is examined. The deuteron two-body photodisintegration and the *ed* elastic scattering reactions are studied. Experimental cross section data are in agreement with the Constituent Counting Rule at relatively low energies, while Hadron Helicity Conservation seems to be not consistent with polarization measurements.

1. Introduction

When a particle strikes a nucleus at low and intermediate energies (or at long and medium distances $d_{NN} > 0.5fm$), the resulting behavior of the nucleus in terms of its constituent nucleons and the mesons that hold them together can be effectively described. The fact that each nucleon is itself made of three quarks held together by gluons is not critical in this regime. When a particle strikes a nucleus at high energies, however, it penetrates the nucleus so deeply that this "effective theory" breaks down, and the nuclear reaction must be described in terms of only quarks and gluons. In this limit the perturbative quantum chromodynamics (pQCD) should be applied.

At present, one of the most interesting problems in nuclear physics is the interplay between these two pictures of the strong interaction: the meson-baryon and the quark-gluon. The main question is at which transfered energy to the nucleon the transition region from hadrons to partons takes place. Exclusive reactions play a central role in studies trying to map out this transition.

2. Exclusive processes and pQCD

Exclusive binary processes are the simplest type of reactions in which to look for the onset of some experimentally accessible phenomena naturally predicted by pQCD. At high energy and high momentum transfer the following two predictions can be derived.

2.1. *The Constituent Counting Rule*

The dimentional scaling [1,2] was derived before the QCD was developed and is based on the simple assumption that at sufficiently high energy $s \to \infty$ and t/s fixed (s and t are the usual Mandelstam variables for the total energy square and the momentum transfer square), hadrons can be described as a collection of free massless quarks with equal momenta. Based on this assumption, simple scaling laws for the scattering amplitudes of a generic binary reactions $AB \to CD$ can be derived. The Constituent Counting Rule (CCR) for the differential cross section can be written as

$$\frac{d\sigma}{dt} = \frac{1}{s^{n-2}} f(\theta_{cm}) \tag{1}$$

where n is the total number of pointlike particles and gauge fields in the initial plus final states.

The validity of this relation has been tested in several hadronic and electromagnetic reactions, like for axample pp scattering or pion photoproduction on nucleon. In the following, we will apply the CCR to study the energy behavior of the deuteron photodisintegration cross section and to the deuteron Form Factors in ed elastic scattering.

2.2. *The Hadron Helicity Conservation*

Based on the same assumption made for CCR, Hadron Helicity Conservation (HHC) can also be derived [3]. In fact, if quark masses can be neglected, the photon-quark and gluon-quark interactions conserve chirality, thus the total helicity between initial and final state must be the same.

So far, no exclusive reactions satisfying HHC have been measured, but the number of studies is limited. In pp elastic scattering the failure of HHC has been attributed to long distance phenomena in which there are three independent scattering of quarks in the beam proton with the quarks in the target proton [4,5]. In photoreactions this mechanism is suppressed, thus HHC is expected to hold [6]. Again, we will examine the HHC law in the case of deuteron photodisintegration and in ed scattering.

3. Deuteron photodisintegration

According to eq. (1), the differential cross section of the deuteron two-body photodisintegration at fixed proton center-of-mass angle should scale following the power law $d\sigma/dt \propto s^{-11}$. In fact, there are six elementary fields for the initial and final nucleons and one for the initial photon, thus $n = 13$.

The first study of this reaction has been performed at SLAC [7], where the differential crosss section at the proton scattering angle $\theta_{cm} = 90^o$ has been found to follow the CCR scaling law already for photon energy above $E_\gamma \approx 1$ GeV. This surprising result led to an extensive program of study of this reaction, first at SLAC [8,9] and then, over a broad range of energy and angles, at JLab [10,11,12,13]. This large amount of data allowed a detailed analysis [14] of the behavior of $d\sigma/dt$ at fixed proton angle as a function of the center-of-mass proton transverse momentum

$$P_T = \sqrt{\frac{1}{2}E_\gamma M_d sin^2(\theta_{cm})} \tag{2}$$

where M_d is the deuteron mass. In fact, P_T is the correct kinematic variable for determine the onset of scaling [15,16]. The experimental data have been fitted with the power law function As^{-11} over window of increasing P_T and the χ^2 of the fits as a function of the minimum P_T of the interval has been studied, applying a statistical criterion to set the threshold for the scaling. It has been found that for the proton angles between 50° and 130° the power law scaling is reached for $P_T \gtrsim 1.1$ GeV/c (with uncertainty of the order of 100 MeV/c). The result of the fits for $P_T > 1.1$ GeV/c are shown in the left panel of Figure 1. For the most forward and backward angles, there are not enough data above $P_T = 1.1$ GeV/c, but the last 3 data points at $\theta_{cm} = 35^o$ show a clear s^{-11} behavior. Polarization observables in deuteron photodisintegration have been much less investigated. In fact, above photon energy of ≈ 1 GeV there are measurements of proton polarization tranfer [17] and of photon polarization asymmetry [18] at $\theta_{cm} = 90^o$ only, as shown in the right panel of Figure 1. These data, in particular the transverse C_x and longitudinal C_z proton polarizations, seems to indicate the HHC could be violated, but more data, at higher energy and other angles, are needed to allow an analysis of the same level of accuracy as done for the cross section.

4. Elastic *ed* scattering

The elastic *ed* scattering is fully described by the electric $G_E(Q^2)$, magnetic $G_M(Q^2)$ and quadrupole $G_Q(Q^2)$ Form Factors, where $Q^2 = -q^2$ and q is the electron momentum transfer. Since the cross section can be written in terms of only two structure functions $A(Q^2)$ and $B(Q^2)$, polarization data are necessary to extract the Form Factors. The deuteron Form Factors have

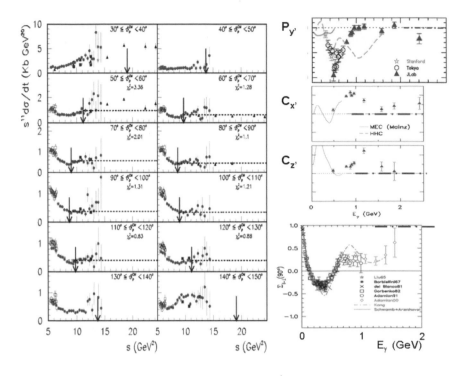

Figure 1. Left side: deuteron two-body differential cross section data at fixed proton center-of-mass angle as a function of the total energy square; the dashed lines are fits of the data with the CCR power law As^{-11} and the arrows indicate the threshold for the scaling (note that $d\sigma/dt$ data are multiplied by s^{11}). Right side: proton polarization transfer (first three plots) and photon polarization asymmetry (fourth plot) as measured in deuteron photodisintegration; the dashed lines indicate the HHC limit.

been consistently extracted [19] in the combined analysis of cross section and tensor polarization t_{20} data, even if some discrepancy between the measurements from different experiments have been found.

The high energy cross section measurement performed at JLab [20] show that the deuteron Form Factor $F_d = \sqrt{A}$ is consistent with the CCR scaling $F_d \propto Q^{-10}$ for Q^2 above ≈ 4 GeV^2, as shown in the left plot of Figure 2.

As for the deuteron photodisintegration, the *ed* polarization data are much less accurate and no definite conclusions can be drawn. The three tensor polarization t_{22}, t_{21} and t_{20} have been measured at JLab [21] for the electron scattering angle $\theta_{cm} = 70^o$ and Q^2 up to ≈ 2 GeV^2 (right plot of Figure 2). While the first two polarizations are not in disagreement with HHC predictions [22] for $Q^2 \gtrsim 1.5$ GeV^2, the trend of the t_{20} data is not consistent with HHC. However, also in this case more data at higher energy and different angles are needed.

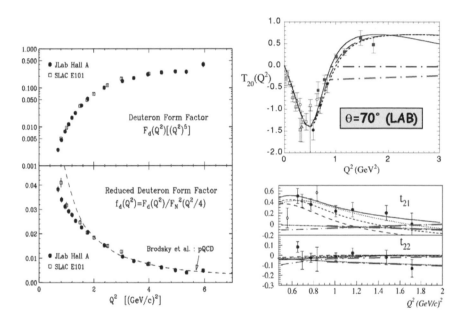

Figure 2. Left side: deuteron Form Factor (multiplied by Q^{10}) as a function of Q^2. Right side: tensor deuteron polarization data measured for electron scattering angle $\theta_{LAB} = 70^o$; the two dot-dashed curves are HHC predictions.

5. Conclusions

The large amount of data on deuteron study with electromagnetic probes showed a general agreement with the CCR at relatively low energies. This is in some way an unexpected result, because the derivation of CCR assumes s and t large compared to all the masses involved in the reaction. However, it must be noted that more realistic QCD calculations could give non negligible corrections to the expected scaling in the few GeV energy region. For example, the recent measurements of proton electromagnetic Form Factors performed at JLab [23] show the scaling law $F_2/F_1 \propto Q^{-1}$ for $Q^2 = $ 4-6 GeV^2, in contrast with the CCR prediction $F_2/F_1 \propto Q^{-2}$. This result has been explained [24] with the contribution of quark orbital angular momentum to the Form Factors, that breaks the CCR scaling law.

HHC is less successful in describing polarization deuteron (and also other hadron-hadron reaction) data in the few GeV energy region. In fact, polarization observables are expected to be more sensitive to the details of the QCD interaction, thus corrections could be large. However, experimental data cover only a limited range of energies and scattering angles, thus more data are necessary before more definite conclusions could be drawn.

References

1. V. Matveev *et al.*, Lett. Nuovo Cimento **7**, 719 (1973).
2. S. L. Brodsky, G. L. Farrar, Phys. Rev. Lett. **31**, 1153 (1973).
3. S. L. Brodsky, G. L. Farrar, Phys. Rev. **D11**, 1309 (1975).
4. T. Gousset *et al.* - Phys. Rev. **D53** (1996) 1202;
5. C. Carlson, M. Chachkhunashvili - Phys. Rev. **D45** (1992) 2555.
6. A. Afanasev *et al.* - Phys. Rev. **D61** (2000) 034014.
7. J. Napolitano *et al.*, Phys. Rev. Lett. **61**, 2530 (1988).
8. S.J. Freedman *et al.*, Phys. Rev. **C48**, 1864 (1993).
9. J.E. Belz *et al.*, Phys. Rev. Lett. **74**, 646 (1995).
10. C. Bochna *et al.*, Phys. Rev. Lett. **81**, 4576 (1998).
11. E.C. Schulte *et al.*, Phys. Rev. Lett. **87**, 102302 (2001).
12. E.C. Schulte *et al.*, Phys. Rev. **C66**, 042201 (2002).
13. M. Mirazita *et al.*, Phys. Rev. **C70**, 014005 (2004).
14. P. Rossi *et al.*, Phys. Rev. Lett. **94**, 012301 (2005).
15. S. J. Brodsky and J. R. Hiller, Phys. Rev. **C28**, 475 (1983).
16. C. E. Carlson, J. R. Hiller, and R. J. Holt, Ann. Rev. Nucl. Part. Sci. **47**, 395 (1997).
17. K. Wijesooriya *et al.*, Phys. Rev. Lett. **86**, 2975 (2001).
18. F. Adamian *et al.* - Eur. Phys. J. **A8** (2000) 423.
19. D. Abbott *et al.*, Eur. Phys. J. **A7** (2000) 421.
20. D. Abbott *et al.*, Phys. Rev. Lett. **82** (1999) 1374.
21. D. Abbott *et al.*, Phys. Rev. Lett. **84** (2000) 5053.
22. S.J. Brodsky and J.R. Hiller, Phys. Rev. **D46** (1992) 2141.
    A. Kobushkin and A. Syamtomov, Phys. Rev. **D49** (1994) 1637.
23. O. Gayou *et al.*, PRL **88**, 092301 (2002).
24. J. Ralston and P. Jain, Phys. Rev. **D69**, 053008 (2004).

EXCLUSIVE BARYON-ANTIBARYON PRODUCTION IN $\gamma\gamma$ COLLISIONS AT e$^+$e$^-$ COLLIDERS

T. BARILLARI

Max-Planck-Inst. für Physik,
Werner-Heisenberg-Institut,
Föhringer Ring 6,
D-80805 München, Germany
E-mail: barilla@mppmu.mpg.de

The exclusive production of baryon-antibaryon pairs in the collisions of two quasi-real photons has been studied using different detectors at e$^+$e$^-$ colliders. Results are presented for $\gamma\gamma \to$ p$\bar{\text{p}}$, $\gamma\gamma \to \Lambda\bar{\Lambda}$, and $\gamma\gamma \to \Sigma^0\bar{\Sigma^0}$ final states. The cross-section measurements are compared with all the existing experimental data and with the analytic calculations based on the three-quark model, on the quark-diquark model, and on the handbag model.

1. Introduction

The exclusive production of baryon-antybaryon (B$\bar{\text{B}}$) pairs in the collision of two quasi-real photons can be used to test predictions of QCD. At e$^+$e$^-$ colliders the photons are emitted by the beam electrons[a] and the B$\bar{\text{B}}$ pairs are produced in the process e$^+$e$^-$ $\to$ e$^+$e$^-\gamma\gamma \to$ e$^+$e$^-$B$\bar{\text{B}}$.

The application of QCD to exclusive photon-photon reactions is based on the work of Brodsky and Lepage [1]. According to their formalism the process is factorized into a non-perturbative part, which is the hadronic wave function of the final state, and a perturbative part. Calculations based on this ansatz [2,3] yields e.g. e$^+$e$^-$ $\to$ e$^+$e$^-\gamma\gamma \to$ e$^+$e$^-$p$\bar{\text{p}}$ cross-sections about one order of magnitude smaller than the existing experimental results [4,5,6,7,8,9], for p$\bar{\text{p}}$ centre-of-mass energies W greater than 2.5 GeV.

Recent studies [10] have extended the systematic investigation of hard exclusive reactions within the quark-diquark model to photon-photon processes [11,12]. In addition, the handbag contribution [13] has been recently proposed to describe the photon-photon annihilation into baryon-antibaryon pairs at large momentum transfer.

---

[a]In this paper positrons are also referred to as electrons.

In this paper, all the existing measurements of the cross-sections for the exclusive $e^+e^- \to e^+e^-B\bar{B}$ processes are presented. In particular, results for $\gamma\gamma \to p\bar{p}$, $\gamma\gamma \to \Lambda\bar{\Lambda}$, and $\gamma\gamma \to \Sigma^0\bar{\Sigma^0}$ final states are reported. These cross-section measurements are compared with the analytic calculations based on the three-quark model, on the quark-diquark model, and on the handbag model.

2. The $\gamma\gamma \to p\bar{p}$ cross-section measurements

The differential cross-section for the process $e^+e^- \to e^+e^-p\bar{p}$ is given by

$$\frac{d^2\sigma(e^+e^- \to e^+e^-p\bar{p})}{dW\, d|\cos\theta^*|} = \frac{N_{ev}(W, |\cos\theta^*|)}{\mathcal{L}_{e^+e^-}\,\varepsilon_{TRIG}\,\varepsilon_{DET}\,(W, |\cos\theta^*|)\,\Delta W\,\Delta|\cos\theta^*|}$$

where N_{ev} is the number of events selected in each $(W, |\cos\theta^*|)$ bin, ε_{TRIG} is the trigger efficiency, ε_{DET} is the detection efficiency, $\mathcal{L}_{e^+e^-}$ is the measured integrated luminosity, and ΔW and $\Delta|\cos\theta^*|$ are the bin widths in W and in $|\cos\theta^*|$.

The total cross-section $\sigma(\gamma\gamma \to p\bar{p})$ for a given value of $\sqrt{s_{ee}}$ is derived from the differential cross-section $d\sigma(e^+e^- \to e^+e^-p\bar{p})/dW$ by using the luminosity function $d\mathcal{L}_{\gamma\gamma}/dW$ [14].

The resulting differential cross-sections for the process $\gamma\gamma \to p\bar{p}$ in bins of W and $|\cos\theta^*|$ are then summed over $|\cos\theta^*|$ to obtain the total cross-section as a function of W for $|\cos\theta^*| < 0.6$.

Fig. 1a) shows the cross-section $\sigma(\gamma\gamma \to p\bar{p})$ measurements as a function of W for $|\cos\theta^*| < 0.6$ obtained by ARGUS [4], CLEO [5], VENUS [6], OPAL [7], L3 [8], and BELLE [9]. Some predictions based on the quark-diquark model [10,11], and the three-quark model [2] are also shown in this figure. There is good agreement between the different experiments results for W > 2.3GeV. At W < 2.3 GeV the OPAL [7] measurements agree with the ARGUS [4] results, but both these measurements lie below the results obtained by CLEO [5], VENUS [6], L3 [8], and BELLE [9].

Within the estimated theoretical uncertainties and for W > 2.2GeV there is a good agreement between the L3 [8] and OPAL [7] results and the quark-diquark model predictions [10,11]. The three-quark model is excluded [2]. At low W the BELLE [9] results are above the quark-diquark model predictions. This measurement agrees with the quark-diquark model for $2.5\,\text{GeV} < W < 3.0\,\text{GeV}$, while at higher W a steeper fall of the BELLE [9] cross-section is observed.

An important consequence of the pure quark hard scattering picture is the power law which follows from the dimensional counting

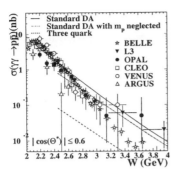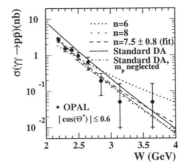

Figure 1. Cross-sections $\sigma(\gamma\gamma \to p\bar{p})$ as a function of W. The data and the theoretical predictions cover a range of $|\cos\theta^*| < 0.6$. a)(Left plot) The experimental data [4,5,6,7,8,9] are compared to the quark-diquark model prediction [10]. The error bars include statistical and systematic uncertainties. b)(Right plot) The data are compared to the quark-diquark model predictions of [11] (dash-dotted line), and of [10] (solid line), using the standard distribution amplitude (DA) with and without neglecting the mass m_p of the proton, and with the predictions of the power law with fixed and with fitted exponent n. The inner error bars are the statistical uncertainties and the outer error bars are the total uncertainties.

rules [15,16]. We expect that for asymptotically large W and fixed $|\cos\theta^*|$, $d\sigma(\gamma\gamma \to p\bar{p})/dt \sim W^{2(2-n)}$ where $n = 8$ is the number of elementary fields and $t = -W^2/2(1 - |\cos\theta^*|)$. The introduction of diquarks modifies the power law by decreasing n to $n = 6$. This power law is compared to the OPAL data in Fig. 1b) with $\sigma(\gamma\gamma \to p\bar{p}) \sim W^{-2(n-3)}$ using three values of the exponent n: fixed values $n = 8$, $n = 6$, and the fitted value $n = 7.5 \pm 0.8$ obtained by taking into account statistical uncertainties only. More data covering a wider range of W would be required to determine the exponent n more precisely.

The measured differential cross-sections $d\sigma(\gamma\gamma \to p\bar{p})/d|\cos\theta^*|$ in different W ranges and for $|\cos\theta^*| < 0.6$ are showed in Fig. 2.

In the range $2.15 < W < 2.55\,\text{GeV}$ the OPAL [7] differential cross-section lies below the results reported by CLEO [5], VENUS [6], L3 [8], and BELLE [9] (Fig. 2a)). Since the CLEO measurements are given for the lower W range $2.0 < W < 2.5\,\text{GeV}$, their results have been rescaled by a factor 0.635 which is the ratio of the two CLEO total cross-section measurements integrated over the W ranges $2.0 < W < 2.5\,\text{GeV}$ and $2.15 < W < 2.55\,\text{GeV}$. This leads to a better agreement between the OPAL and CLEO measurements but the OPAL results are still consistently lower. The shapes of the $|\cos\theta^*|$ dependence of all measurements are consistent.

Fig. 2b) shows the differential cross-sections $d\sigma(\gamma\gamma \to p\bar{p})/d|\cos\theta^*|$ in the W range $2.5 < W < 3.0\,\mathrm{GeV}$ obtained by CLEO [5], OPAL [7], L3 [8], and BELLE [9] in similar W ranges, these differential cross-section have been normalized to the that averaged within $|\cos\theta^*| < 0.3$. The measurements are consistent within the uncertainties.

The comparison of the differential cross-section as a function of $|\cos\theta^*|$ for $2.55 < W < 2.95\,\mathrm{GeV}$ with the calculation of [10] at $W = 2.8\,\mathrm{GeV}$ for different distribution amplitudes (DA) is also shown in this figure together with pure quark model [2] and the handbag model prediction [13]. The shapes of the curves are consistent with those of the data. Fig. 2 shows that

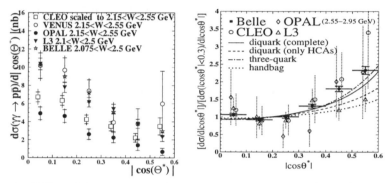

Figure 2. Differential cross-sections for $\gamma\gamma \to \mathrm{p\bar{p}}$ as a function of $|\cos\theta^*|$ in different ranges of W; a) (left plot) low range $2.15 < W < 2.55\,\mathrm{GeV}$, (b) (right plot) higher range $2.5 < W < 3.0\,\mathrm{GeV}$. The inner error bars are the statistical uncertainties and the outer error bars are the total uncertainties.

the differential cross-section at low W decreases at large $|\cos\theta^*|$, while the opposite trend is observed in the higher W region. The transition point seems to occur at $W \approx 2.5\,\mathrm{GeV}$ [9].

Another important consequence of the hard scattering picture is the hadron helicity conservation rule. For each exclusive reaction like $\gamma\gamma \to \mathrm{p\bar{p}}$ the sum of the two initial helicities equals the sum of the two final ones. According to the simplification used in [11], neglecting quark masses, quark and antiquark and hence proton and antiproton have to be in opposite helicity states. If the (anti) proton is considered as a point-like particle, simple QED rules determine the angular dependence of the unpolarized $\gamma\gamma \to \mathrm{p\bar{p}}$ differential cross-section [17]:

$$\frac{\mathrm{d}\sigma(\gamma\gamma \to \mathrm{p\bar{p}})}{\mathrm{d}|\cos\theta^*|} \propto \frac{(1 + \cos^2\theta^*)}{(1 - \cos^2\theta^*)}. \tag{1}$$

This expression is compared to the OPAL [7] data in two W ranges, $2.55 < W < 2.95\,\mathrm{GeV}$ (Fig. 3a) and $2.15 < W < 2.55\,\mathrm{GeV}$ (Fig. 3b).

The normalisation in each case is determined by the best fit to the data. In the higher W range, the prediction (1) is in agreement with the data within the experimental uncertainties. In the lower W range this simple model does not describe the data. At low W soft processes such as meson exchange are expected to introduce other partial waves, so that the approximations leading to (1) become invalid.

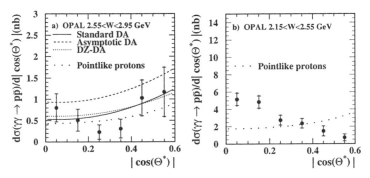

Figure 3. Measured differential cross-section, $d\sigma(\gamma\gamma \to p\bar{p})/d|\cos\theta^*|$, with statistical (inner bars) and total uncertainties (outer bars) for a) $2.55 < W < 2.95$ GeV and b) $2.15 < W < 2.55$ GeV. The data are compared with the point-like approximation for the proton (1) scaled to fit the data. The other curves show the pure quark model [2], the diquark model of [11] with the Dziembowski distribution amplitudes (DZ-DA), and the diquark model of [10] using standard and asymptotic distribution amplitudes.

3. The $\gamma\gamma \to \Lambda\overline{\Lambda}$ and $\gamma\gamma \to \Sigma^0\overline{\Sigma^0}$ cross-section measurements

The cross-sections $\sigma(\gamma\gamma \to \Lambda\overline{\Lambda})$ and $\sigma(\gamma\gamma \to \Sigma^0\overline{\Sigma^0})$ in real photon collisions as a function of W and for $|\cos\theta^*| < 0.6$ can be extracted by deconvoluting the two-photon luminosity function and the form factor [14].

Fig. 4 compares the L3 [18] $\sigma(\gamma\gamma \to \Lambda\overline{\Lambda})$ measurement with that obtained by CLEO [19]. For $W > 2.5$ GeV the two results are compatible inside the large experimental errors. The cross-section measurement obtained by CLEO at lower W values is steeper that the one obtained by L3. The L3 [18] data, fitted with a function of the form $\sigma \approx W^{-n}$, gives a value $n = 7.6 \pm 3.9$. In Fig. 4 the $\sigma(\gamma\gamma \to \Lambda\overline{\Lambda})$ and $\sigma(\gamma\gamma \to \Sigma^0\overline{\Sigma^0})$ cross-section measurements are compared to the predictions of the quark-diquark model calculation [20]. The absolute predictions using the standard distribution amplitude [20] (Standard DA) reproduce well the L3 data, the asymptotic DA and the DZ-DA models [20] are excluded. The CLEO [19] and L3 [18] $\sigma(\gamma\gamma \to \Lambda\overline{\Lambda})$ cross-section measurements and L3 $\sigma(\gamma\gamma \to \Sigma^0\overline{\Sigma^0})$ cross-section measurements for $W > 2.5\,\mathrm{GeV}$ are satisfactory described also by the handbag model, see Ref. [13].

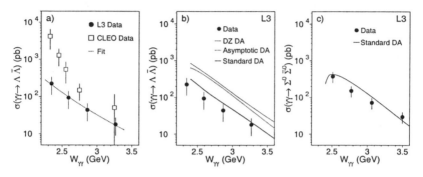

Figure 4. Measurements of the $\sigma(\gamma\gamma \to \Lambda\overline{\Lambda})$ and $\sigma(\gamma\gamma \to \Sigma^0\overline{\Sigma^0})$ cross-sections as a function of W. In a) the $\sigma(\gamma\gamma \to \Lambda\overline{\Lambda})$ cross-section is compared to the one obtained by CLEO [19]. The dashed line shows the power law fit as desccribed in the text. In b) and c) the $\sigma(\gamma\gamma \to \Lambda\overline{\Lambda})$ and $\sigma(\gamma\gamma \to \Sigma^0\overline{\Sigma^0})$ measurements are compared to the quardk-diquark model predictions [20]

References

1. G.P. Lepage and S.J. Brodsky, Phys. Rev. **D22** 2157 (1980).
2. G.R. Farrar, E. Maina and F. Neri, Nucl. Phys. **B259** 702 (1985).
3. V.L. Chernyak and I.R. Zhitnitsky, Nucl. Phys. **B246** 52 (1984).
4. ARGUS Collaboration, H. Albrecht et al., Z. Phys. **C42** 543 (1989).
5. CLEO Collaboration, M. Artuso et al., Phys. Rev. **D50** 5484 (1994).
6. VENUS Collaboration, H. Hamasaki et al., Phys. Lett. **B407** 185 (1997).
7. OPAL Collaboration, G. Abbiendi et al., Eur. Phys. J. **C28** 45 (2003).
8. L3 Collaboration, P. Achard et al., Phys. Lett. **B571** 11 (2003).
9. BELLE Collaboration, Chen-Cheng Kuo et al., Phys. Lett. **B621** 41 (2005).
10. C.F. Berger and W. Schweiger, Eur. Phys. J. **C28** 249 (2003).
11. M. Anselmino, F. Caruso, P. Kroll and W. Schweiger, Int. J. Mod. Phys. **A4** 5213 (1989).
12. P. Kroll, M. Schürmann and W. Schweiger, Int. J. Mod. Phys. **A6** 4107 (1991).
13. M. Diehl, P. Kroll and G. Vogt, Eur. Phys. J. **C26** 567 (2003).
14. G. A. Schuler, Comp. Phys. Comm. **108** 279 (1998).
15. S.J. Brodsky and G.R. Farrar, Phys. Rev. Lett. **31** 1153 (1973).
16. V.A. Matveev, R.M. Muradian and A.N. Tavkhelidze, Nuovo Cim. Lett. **7** 719 (1973).
17. V.M. Budnev, I.F. Ginzburg, G.V. Meledin and V.G. Serbo, Phys. Rep. **15** 181 (1974).
18. L3 Collaboration, P. Achard et al., Phys. Lett. **B536** 24 (2002).
19. CLEO Collaboration, S. Anderson et al., Phys. Rev. **D56** 2485 (1997).
20. C.F. Berger and W. Schweiger, Eur. Phys. J. **C39** 173 (2005).

PHOTOABSORPTION AND PHOTOPRODUCTION ON NUCLEI IN THE RESONANCE REGION

S. SCHADMAND

Institut für Kernhysik
Forschungszentrum Jülich
D-52425 Jülich, Germany
E-mail: s.schadmand@fz-juelich.de

Inclusive studies of nuclear photoabsorption have provided clear evidence of medium modifications in the properties of hadrons. However, the results have not been explained in a model independent way. A deeper understanding of the situation is expected from a detailed comparison of meson photoproduction from nucleons and from nuclei in exclusive reactions. Recent experimental results are presented.

1. Introduction

Current issues in the understanding of the strong interaction address the structure of hadrons, consisting of quarks and gluons, as the building blocks of matter. Central challenges concern the questions why quarks are confined within hadrons and how hadrons are constructed from their constituents. One goal is to find the connection between the parton degrees of freedom and the low energy structure of hadrons leading to the study of the hadron excitation spectrum and the search for exotic states, like glueballs or hybrid states. An approach related to the question of the origin of hadron masses is the search for modifications of hadron properties in the nuclear medium. The underlying question is the origin of hadron masses in the context of chiral symmetry breaking. Evidence for such effects has been searched for in many experiments. In this contribution, photoabsorption and meson photoproduction on nuclei are discussed.

2. Nuclear Photoabsorption

Photoabsorption experiments on the free nucleon demonstrate the complex structure of the nucleon and its excitation spectrum, as shown in Fig. 1. The lowest-lying peak is the $\Delta(1232)$ resonance and is prominently excited by incident photons of 0.2–0.5 GeV. The following group of resonances, $P_{11}(1440)$, $D_{13}(1520)$, and $S_{11}(1535)$, is called the second resonance region (E_γ=0.5-0.9 GeV), a third resonance region is visible. The observed resonance structures have been studied using their decay via light mesons, showing that the photoabsorption spectrum can be explained by the sum of π, $\pi\pi$ and η production cross sections.

Fig. 1 also shows the nuclear photoabsorption cross section per nucleon as an average over the nuclear systematics [1]. The Δ resonance is broadened

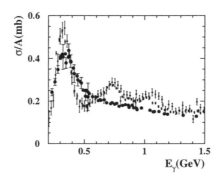

Figure 1.   Nuclear photoabsorption cross section per nucleon as an average over the nuclear systematics [1] (full symbols) compared to the absorption on the proton [2] (open symbols).

and slightly shifted while the second and higher resonance regions seem to have disappeared.

Mosel et al. [3,4,5] have argued that an in-medium broadening of the $D_{13}(1520)$ resonance is a likely cause of the suppressed photoabsorption cross section. Hirata et al. [6] see a change of the interference effects in the nuclear medium as one of the most important reasons for the suppression of the resonance structure. However, the absence of resonance structure in nuclear photoabsorption has not been explained in a model-independent way. A deeper understanding of the situation is anticipated from the experimental study of meson photoproduction on nucleons embedded in nuclei in comparison to studies on the free nucleon.

3. Meson Production in the Second Resonance Region

In the second resonance region, double pion production aims at the resonances $D_{13}(1520)$ and $P_{11}(1440)$ while η production is characteristic for the $S_{11}(1535)$ resonance. The three resonances in the second resonance region decay to roughly 50% via single pion emission. The most trivial medium modification is the broadening of the excitation functions due to Fermi motion. The decay of the resonances is further modified by Pauli-blocking of final states, which reduces the resonance widths. In addition, decay channels like $N^\star N \to NN$ cause collisional broadening. Both effects could cancel to some extent.

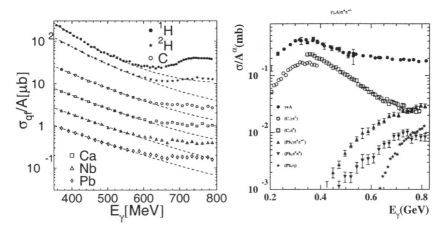

Figure 2. Left: Total cross section per nucleon for single π° photoproduction in the second resonance region for the nucleon and for nuclei. The scale corresponds to the proton data, the other data are scaled down by factors 2,4,8,16,32, respectively. The dashed curves are fits to the data in the energy range 350–550 MeV. Right: Status of the decomposition of nuclear photoabsorption into meson production channels (scaled with A^α, $\alpha=2/3$). Small open circles are the average nuclear photoabsorption cross section per nucleon ($\alpha=1$) [1]. Meson production data are from [7,8,9,10,11,12].

An attempt to study the in-medium properties of the D_{13} resonance was undertaken with a measurement of quasifree single π° photoproduction [8] which, on the free nucleon, is almost exclusively sensitive to the D_{13} resonance. The left panel of Fig. 2 summarizes the results. Strong quenching of the D_{13}-resonance structure is found for the deuteron with respect to the nucleon. However, an indication of a broadening or a suppression of the D_{13} structure in heavy nuclei is not observed. Model predictions agree with the pion photoproduction data only under the assumption of a strong broadening of the resonance, other effects seem to be missing in the models. This casts doubt on the interpretation of the total photoabsorption data via resonance broadening. In contrast to the case of total photoabsorption, the second resonance bump remains visible. However, exclusive reaction channels are dominated by the nuclear surface region where in-medium effects are smaller. Furthermore, as discussed in [13], resonance broadening effects are even more diluted for reactions which do not contribute to the broadening, due to the averaging over the nuclear volume.

The right panel of Fig. 2 shows the status of the decomposition of nuclear photoabsorption into meson production channels. The available experimental meson cross sections are exclusive measurements, investigating quasifree production. The purely charged final states have not been measured. However, it can be inferred from the existing data that the sum of the cross sections would not reproduce the flat shape of the total photoabsorption from nuclei. In a recent compilation [14], it is observed that the current results indicate large differences between quasifree meson production from the nuclear surface and non-quasifree components. The quasifree part does not show a suppression of the resonance structures in the second resonance region. However, resonance structures seem absent in the non-quasifree meson production which has larger contributions from the nuclear volume.

4. ω Mesons in the nuclear Medium

The photoproduction of ω mesons on nuclei has been investigated using the Crystal Barrel/TAPS experiment at the ELSA tagged photon facility in Bonn [15]. The aim is to study possible in-medium modifications of the ω meson via the reaction $\gamma + A \rightarrow \omega + X \rightarrow \pi^\circ\gamma + X'$. A number of theoretical models predict a mass shift of the omega meson in the nuclear medium, for references see [16,15]. An excellent recent review discusses nucleon and hadron structure changes in the nuclear medium within the quark-meson coupling (QMC) model predicting a reduction in the omega mass in nuclei [17]. Experimentally, results obtained for Nb are compared to a reference measurement on a LH_2 target. The left panel of Fig. 3 shows the $\pi^\circ\gamma$ invariant mass distribution without further cuts except for a three momentum cutoff of $|\vec{p}_\omega| < 500$ MeV/c. The dominant background source is two pion production where one of the four photons escapes the detection. This probability was determined by Monte Carlo simulations to be 14%. The resulting three photon final state is not distinguishable from the $\omega \rightarrow \pi^\circ\gamma$ invariant mass. These decays are eliminated by matching the right hand part of the Nb invariant mass spectrum to the LH_2 data (see central panel of Fig. 3) and by subtracting the two spectra from each other. For this normalization the integral of the undistorted spectrum corresponds to 75% of

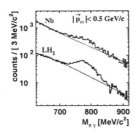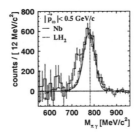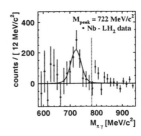

Figure 3. Left panel: Inclusive $\pi^\circ\gamma$ invariant mass spectra for ω momenta less than 500 MeV/c. Upper histogram: Nb data, lower histogram: LH$_2$ target reference measurement. The dashed lines indicate fits to the respective background. Middle panel: $\pi^\circ\gamma$ invariant mass for the Nb data (solid histogram) and LH$_2$ data (dashed histogram) after background subtraction. The error bars show statistical uncertainties only. The solid curve represents the simulated line shape for the LH$_2$ target. Right panel: In-medium decays of ω mesons along with a Voigt fit to the data. The vertical line indicates the vacuum ω mass of 782 MeV/c^2.

the counts in the Nb spectrum. This is in good agreement with a theoretical prediction obtained from a transport code calculation [18,19]. There, about 16% of the total decays are predicted to occur inside the nuclear medium ($\rho > 0.1 \cdot \rho_0$) without final state interaction (FSI) and 3% of the events are distorted due to FSI in the mass range of $0.6\,\text{GeV/c}^2 < M_{\pi^\circ\gamma} < 0.9\,\text{GeV/c}^2$. In addition, 9% of the events are moved towards lower masses due to the Δ decay kinematics. The right panel of Fig. 3 shows the invariant mass distribution obtained after background subtraction. The expected superposition of decays outside of the nucleus is observed at the nominal vacuum mass with decays occurring inside the nucleus, responsible for the shoulder towards lower invariant masses. The high mass part of the ω mass signal appears to be identical for the Nb and LH$_2$ targets, indicating that this part is dominated by ω meson decays in vacuum.

A difference in the line shape for the two data samples is not observed for recoiling, long-lived mesons (π°, η and η'), which decay outside of the nucleus. However, for ω mesons produced on the Nb target a significant enhancement towards lower masses is found. For momenta less than 500 MeV/c an in-medium ω meson mass of $M_{\text{medium}} = [722^{+2}_{-2}(\text{stat})^{+35}_{-5}(\text{syst})]\,\text{MeV/c}^2$ has been deduced at an estimated average nuclear density of 0.6 ρ_0.

112

5. Summary

The systematic study of total cross sections for single $\pi°$, η, and $\pi\pi$ production over a series of nuclei has not provided an obvious hint for a depletion of resonance yield. The observed reduction and change of shape in the second resonance region are mostly as expected from absorption effects, Fermi smearing and Pauli blocking, and collisional broadening. The sum of experimental meson cross sections for neutral and mixed charged states between 400 and 800 MeV demonstrates the persistence of the second resonance bump when at least one neutral meson is observed. It has to be concluded that the medium modifications leading to the depletion of cross section in nuclear photoabsorption are a subtle interplay of effects. Their investigation and the rigorous comparison to theoretical models requires the detailed study of differential cross sections and a deeper understanding of meson production in the nuclear medium. The recent experimental investigation of ω photoproduction from nuclei has been presented as one detailed study of medium modification of hadrons. First evidence for a lowering of the ω mass in the nuclear medium has been observed.

References

1. V. Muccifora, et al., Phys. Rev. C60 (1999) 064616.
2. K. Hagiwara, et al., Phys. Rev. D66 (2002) 010001.
3. U. Mosel, Prog. Part. Nucl. Phys. 42 (1999) 163–176.
4. J. Lehr, M. Effenberger, U. Mosel, Nucl. Phys. A671 (2000) 503–531.
5. M. Effenberger, A. Hombach, S. Teis, U. Mosel, Nucl. Phys. A614 (1997) 501–520.
6. M. Hirata, N. Katagiri, K. Ochi, T. Takaki, Phys. Rev. C66 (2002) 014612.
7. J. Arends, et al., Z. Phys. A305 (1982) 205.
8. B. Krusche, et al., Phys. Rev. Lett. 86 (2001) 4764–4767.
9. M. Roebig-Landau, et al., Phys. Lett. B373 (1996) 45–50.
10. H. Yamazaki, et al., Nucl. Phys. A670 (2000) 202–205.
11. S. Janssen, PhD thesis, University of Giessen (2002).
12. S. Schadmand, Habilitation thesis, University of Giessen (2005).
13. J. Lehr, U. Mosel, Phys. Rev. C64 (2001) 042202.
14. B. Krusche, et al., Eur. Phys. J. A22 (2004) 347–351.
15. D. Trnka, et al., Phys. Rev. Lett. 94 (2005) 192303.
16. J. G. Messchendorp, A. Sibirtsev, W. Cassing, V. Metag, S. Schadmand, Eur. Phys. J. A11 (2001) 95–103.
17. K. Saito, K. Tsushima, A. W. Thomas, hep-ph/0506314.
18. P. Muhlich, T. Falter, U. Mosel, Eur. Phys. J. A20 (2004) 499–508.
19. P. Muhlich, U. Mosel, private ecommunication.

Duality in Nuclei

A PARTONIC PICTURE OF JET FRAGMENTATION IN NUCLEI

XIN-NIAN WANG

Nuclear Science Division, MS 70R0319, Lawrence Berkeley National Laboratory
Berkeley, California 94720
E-mail: xnwang@lbl.gov

Inclusive hadron production in deeply inelastic scattering off a large nucleus (DISA) involves parton propagation through the nuclear medium and hadronization which could start inside the nucleus. In most of the kinematic region accessible in high-energy DISA, the hadron formation time is much larger than the nuclear size. Therefore, leading hadron suppression in DISA can be described by modification of the parton fragmentation via multiple scattering and induced energy loss.

1. Introduction

In QCD, quarks are the basic fundamental building blocks for various hadrons. Yet, in the context of quark-hadron duality in the study of strong interaction, both can be used to describe the same phenomenon around the crossover of quark and hadron degrees of freedom, since the complex nonperturbative quark interaction can be be effectively described by hadronic interaction at low energy scales. However, with increasing energy scale, eventually quarks have to become the only relevant degrees of freedom.

In the study of leading hadron production in deeply inelastic scattering off a large nucleus (DISA), the relevant (or more effective) degrees of freedom are quarks during the short-distance photon-nucleus interaction. The knocked-out quark will inevitably have to propagate through at least part of the nucleus before hadronization. If the hadronization happens inside the nucleus, one apparently should resort to effective hadron-hadron interaction to describe the re-scattering between hadrons from the quark fragmentation and nucleons inside the nucleus [1].

In a simple picture of hadronization of a quark jet in high-energy DISA, one can consider that it starts with the radiation of a virtual gluon which subsequently splits into a pair of quark and anti-quark. The anti-quark has a finite probability to form a color singlet dipole with the leading quark [2], which could become a final leading hadron. The hadron formation time (time for the build-up of the wavefunction), $t_h \sim r_h E/m$, could be much larger than the nuclear size R_A, where r_h is the hadron size and m the hadron mass. It is therefore a good assumption that physical hadrons are formed outside the nucleus. The formation time of the dipole or the so-called pre-hadron can be estimated as the formation time of the virtual gluon,

$$\tau_f = 2Ez_g(1 - z_g)/k_T^2 \tag{1}$$

which can also be larger than the nuclear size for moderately large z_g if one takes $k_T \sim \Lambda_{\text{QCD}}$. Therefore, most of the pre-hadrons are also formed outside the nucleus. Hadron production as a result of the collinear radiation (or leading-log approximation) will eventually fall off at large z_h as $(1-z_h)^n$. For $z_h \sim 1$, exclusive processes will become dominate even though they are suppressed by $1/Q^4$ for each final hadron state as it is produced at a short distance $\sim 1/Q$.

During the final state DGLAP radiation, even though the average total duration,

$$\langle \tau_d \rangle = \int_{z_0}^{1-z_0} dz \int_{Q_0^2}^{z(1-z)Q^2} \frac{2Ez(1-z)}{\ell_T^2} \frac{dN_g}{dz d\ell_T^2} \approx \frac{\alpha_s}{\pi} C_F \frac{2E}{Q_0^2} \tag{2}$$

with $z_0 = Q_0^2/Q^2$ and $Q_0 \sim 200$ MeV, might be long, the average duration for the first few radiations can still be very short. Therefore, the DGLAP evolution and induced radiation due to final state multiple scattering should be considered together in the same framework of jet fragmentation, while one can neglect hadronic interaction inside the nucleus.

2. Modified jet fragmentation

The differential semi-inclusive hadronic tensor of DISA in a collinear factorization approximation to the leading twist can be written as

$$\frac{dW_{\mu\nu}^S}{dz_h} = \sum_q \int dx f_q^A(x, \mu_I^2) H_{\mu\nu}(x, p, q) D_{q \to h}(z_h, \mu^2) , \tag{3}$$

in an infinite momentum frame, where the photon carries momentum $q = [-x_B p^+, q^-, \vec{0}_\perp]$ and the momentum of the target per nucleon is $p = [p^+, 0, \vec{0}_\perp]$ with the Bjorken variable defined as $x_B = Q^2/2q^- p^+$. $f_q^A(x, \mu_I^2)$ is the quark distribution of the nucleus, $H_{\mu\nu}(x, p, q)$ is the hard part of $\gamma^* + q$ scattering and $D_{q \to h}(z_h, \mu^2)$ is the quark fragmentation function in vacuum.

In a nucleus target, the outgoing quark can scatter again with another parton from the nucleus. The rescattering may induce additional gluon radiation and cause the leading quark to lose energy. Such induced gluon radiation will effectively give rise to additional terms in the evolution equation leading to modification of the fragmentation functions in a medium [3]. Contributions from multiple parton scattering are always non-leading twist. However we will consider only those that are enhanced by the nuclear size $A^{1/3}$.

Within the LQS framework [4] of generalized factorization, one can calculate the radiative correction due to double scattering processes. The final results, together with the leading order contribution, can be expressed as an effective modified fragmentation function [3],

$$\tilde{D}_{q\to h}(z_h, \mu^2) \equiv D_{q\to h}(z_h, \mu^2) + \int_0^{\mu^2} \frac{d\ell_T^2}{\ell_T^2} \frac{\alpha_s}{2\pi} \int_{z_h}^1 \frac{dz}{z}$$
$$\times \left[\Delta\gamma_{q\to qg}(z) D_{q\to h}(\frac{z_h}{z}) + \Delta\gamma_{q\to gq}(z) D_{g\to h}(\frac{z_h}{z}) \right], \quad (4)$$

where $\Delta\gamma_{q\to gq}(z) = \Delta\gamma_{q\to qg}(1-z)$ and,

$$\Delta\gamma_{q\to qg}(z) = \left[\frac{1+z^2}{(1-z)_+} T_{qg}^A(x, x_L) + \delta(1-z)\Delta T_{qg}^A(x, \ell_T^2) \right]$$
$$\times \frac{C_A 2\pi\alpha_s}{(\ell_T^2 + \langle k_T^2 \rangle) N_c f_q^A(x, \mu_I^2)} \quad (5)$$

are the modified splitting functions for the induced gluon radiation. The δ-function part is from the virtual correction contribution with $\Delta T_{qg}^A(x, \ell_T^2)$ defined as

$$\Delta T_{qg}^A(x, \ell_T^2) \equiv \int_0^1 dz \frac{1}{1-z} \left[2T_{qg}^A(x, x_L)|_{z=1} - (1+z^2) T_{qg}^A(x, x_L) \right]. \quad (6)$$

$$T_{qg}^A(x, x_L) = \int \frac{dy^-}{2\pi} dy_1^- dy_2^- e^{i(x+x_L)p^+ y^-} (1 - e^{-ix_L p^+ y_2^-})(1 - e^{-ix_L p^+ (y^- - y_1^-)})$$
$$\tfrac{1}{2} \langle A|\bar{\psi}_q(0) \gamma^+ F_\sigma^{+}(y_2^-) F^{+\sigma}(y_1^-) \psi_q(y^-)|A\rangle \theta(-y_2^-)\theta(y^- - y_1^-) \quad (7)$$

is the quark-gluon correlation function which also contains LPM interference effect. Here $x_L = \ell_\perp^2 / 2p^+ q^- z(1-z)$ and $1/x_L p^+$ represent the gluon formation time.

Assuming a Gaussian nuclear distribution in the rest frame, $\rho(r) \sim \exp(-r^2/2R_A^2)$, $R_A = 1.12A^{1/3}$ fm, one can approximate T_{qg}^A in terms of single parton distributions [5],

$$T_{qg}^A(x, x_L) = \tilde{C} m_N R_A f_q^A(x)(1 - e^{-x_L^2/x_A^2}), \quad (8)$$

where $x_A = 1/m_N R_A$, and m_N is the nucleon's mass. The second term corresponds to LPM interference contributions and involve transferring momentum x_L between different nucleons inside a nucleus and thus should be suppressed for large nuclear size or large momentum fraction x_L. Notice $x_L/x_A = L_A/\tau_f$ with $L_A = R_A m_N/p^+$ being the nuclear size in the infinite momentum frame. Because of the LPM interference effect, the above effective parton correlation and the induced gluon emission vanishes when $x_L/x_A \ll 1$. Therefore, the LPM interference requires the radiated gluon to have a minimum transverse momentum $\ell_T^2 \sim Q^2/M R_A \sim Q^2/A^{1/3}$. The nuclear corrections to the fragmentation function due to double parton scattering will then be in the order of $\alpha_s A^{1/3}/\ell_T^2 \sim \alpha_s A^{2/3}/Q^2$, which depends quadratically on the nuclear size.

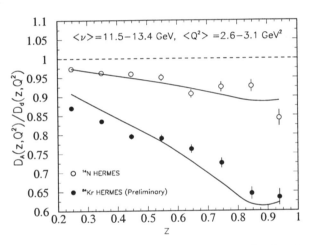

Figure 1.   The predicted nuclear modification of jet fragmentation function is compared to the HERMES data [6].

Shown in Fig. 1 are the calculated nuclear modification factor of the fragmentation function for ^{14}N and ^{84}Kr targets as compared to the HERMES data [6]. The only parameter in our calculation is set $\widetilde{C}\alpha_s^2 = 0.00065$ GeV^2. The predicted shape of the z dependence and the quadratic nuclear size dependence agrees well with the experimental data. The energy dependence of the suppression also has excellent agreement with our prediction [7]. What is amazing is the clear quadratic $A^{2/3}$ nuclear size dependence of the suppression which is truly a non-Abelian effect in QCD.

With the modified fragmentation function in Eq. (4), one can calculate theoretically the average energy loss by the quark, which is the energy carried away by the radiated gluons,

$$\Delta E = \nu \langle \Delta z_g \rangle \approx \tilde{C} \alpha_s^2 m_N R_A^2 (C_A/N_c) 3 \ln(1/2x_B). \tag{9}$$

With the value of $\alpha_s^2 \tilde{C}$, and $L_A = R_A \sqrt{2\pi}$ one gets the quark energy loss $dE/dx \approx 0.5$ GeV/fm for a Au nuclear target.

3. Dihadron fragmentation functions

To further investigate the mechanism of nuclear suppression of leading hadron spectra in DISA, dihadron correlation or fragmentation functions could be a more sensitive measurement. The dihadron fragmentation functions in terms of the overlapping matrix between parton field operators and the final hadron states have been defined and their DGLAP evolution equations have been derived recently [8], which are similar to that of single hadron fragmentation functions. However, there are extra contributions that are proportional to the convolution of two single hadron fragmentation functions. These correspond to independent fragmentation of both daughter partons after the parton split in the radiative processes. Medium modification to the dihadron fragmentation functions due to induced radiation was found to have the identical form as the DGLAP evolution equations [9].

These medium modifications depend on the same gluon correlation functions as in the modification to the single hadron fragmentation functions. Therefore, in the numerical calculation of the medium modification of dihadron fragmentation functions, there are no additional parameters involved. The predicted results for jet quenching in DIS are in good agreement with HERMES data [9] as shown in Fig. 2. The nuclear modification is found to manifest mostly in the single hadron fragmentation functions. Since dihadron fragmentation functions already contain the information of single hadron fragmentation function, the modification to the remaining correlated distribution, $D_q^{h_1 h_2}(z_1, z_2)/D_q^{h_1}(z_1)$ is very small.

Acknowledgement

I would like to thank A. Majumder and E. Wang for their collaboration. This work is supported the Director, Office of Energy Research, Office of High Energy and Nuclear Physics, Division of Nuclear Physics, and by the Office of Basic Energy Science, Division of Nuclear Science, of the U. S. Department of Energy under Contract No. DE-AC02-05CH11231.

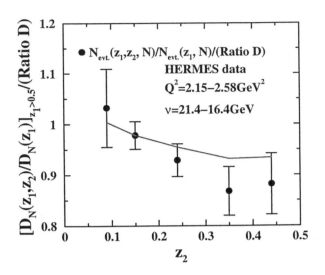

Figure 2. Medium modification of the dihadron distribution (normalized by single hadron distribution) versus its momentum fraction as compared to HERMES data [10].

References

1. T. Falter, W. Cassing, K. Gallmeister and U. Mosel, Phys. Rev. C **70**, 054609 (2004).
2. B. Z. Kopeliovich, L. B. Litov and J. Nemchik, Int. J. Mod. Phys. E **2** (1993) 767.
3. X. F. Guo and X.-N. Wang, Phys. Rev. Lett. **85** (2000) 3591. X. N. Wang and X. F. Guo, Nucl. Phys. A **696**, 788 (2001).
4. M. Luo, J. Qiu and G. Sterman, Phys. Lett. B **279** (1992) 377; Phys. Rev. D **50** (1994) 1951; Phys. Rev. D **49** (1994) 4493.
5. J. Osborne and X. N. Wang, Nucl. Phys. A **710**, 281 (2002).
6. A. Airapetian *et al.* [HERMES Collaboration], Eur. Phys. J. C **20**, 479 (2001).
7. E. Wang and X. N. Wang, Phys. Rev. Lett. **89**, 162301 (2002).
8. A. Majumder and X. N. Wang, Phys. Rev. D **70**, 014007 (2004). A. Majumder and X. N. Wang, Phys. Rev. D **72**, 034007 (2005).
9. A. Majumder, E. Wang and X. N. Wang, arXiv:nucl-th/0412061.
10. [HERMES Collaboration], arXiv:hep-ex/0510030.

HADRON ATTENUATION BY (PRE)HADRONIC FSI AT HERMES

T. FALTER[1], K. GALLMEISTER[2] AND U. MOSEL[2]

[1] *Physics Department, Brookhaven National Laboratory, USA*
[2] *Institut für Theoretische Physik, Universität Gießen, Germany*

We investigate hadron production in deep inelastic lepton-nucleus scattering in the kinematic regime of the HERMES experiment. Our calculations are carried out in the framework of a BUU transport model which contains the Lund event generators PYTHIA and FRITIOF for the simulation of high-energy elementary interactions. For the first time we consistently use the complete four-dimensional information of the Lund string break up vertices as input for our transport theoretical studies of (pre)hadronic final state interactions. We compare our results with experimental HERMES data on charged hadron attenuation.

1. Introduction

Hadron production in deep inelastic lepton-nucleus scattering (DIS) provides an ideal tool to investigate the space-time evolution of hadron formation[1]. The nuclear target can be viewed as a kind of 'micro-detector' that is located directly behind the virtual photon-nucleon interaction vertex. It allows us to study the interactions of the reaction products with the surrounding nuclear environment on a length scale that is set by the size of the target nucleus. By comparison with hadron production on a deuterium target one can draw conclusions about the space-time picture of hadronization. The latter information is crucial for the interpretation of jet quenching in ultra-relativistic heavy ion-collisions at RHIC as a possible signature for the creation of a deconfined quark-gluon plasma phase[2].

In the recent past the HERMES collaboration at DESY has started an extensive experimental investigation of hadron attenuation in DIS on various gas targets[3] and a similar experiment is currently performed at Jefferson Lab[4]. Furthermore, hadron attenuation in nuclear DIS will be subject to investigation after the 12 GeV upgrade at Jefferson Lab and at a possible future electron-ion collider[5]. In Refs. 6 the observed attenuation of high energy hadrons at HERMES has been interpreted as being due to a partonic energy loss of the quark that was struck by the virtual photon. The colored quark undergoes rescattering inside the nucleus giving rise to induced gluon radiation. However, the authors of Refs. 7 and ourselves[8,9] achieve a very good description of the experimental data by assuming that the struck quark forms a color neutral prehadron early after the virtual photon-nucleon interaction. This prehadron then scatters off the surrounding nucleons on its way out of the nucleus.

In Refs. 8, 9 and references therein we have developed a method to incorporate coherence length effects in a semi-classical Boltzmann-Uehling-Uhlenbeck (BUU) transport model that uses the Lund string models PYTHIA[10] and FRITIOF[11] for the simulation of the elementary DIS process and the (pre)hadronic final state interactions (FSI) respectively. This allows for a complete probabilistic coupled-channel description of high-energy photo- and electroproduction off complex nuclei. Our simulation works on an event-by-event basis and can be directly compared to experiment accounting for all sorts of kinematic cuts and detector acceptances. Our theoretical results are in perfect agreement with the experimental findings at HERMES.

2. Model

In our model we split the lepton-nucleus interaction into two parts: In step 1) the exchanged virtual photon interacts with a bound nucleon inside the nucleus and produces a final state that in step 2) is propagated within the transport model. We take into account nuclear effects such as binding energies, Fermi motion and Pauli blocking. We also account for shadowing of the resolved photon components using the method developed in Refs. 12.

The final state of the initial photon-nucleon reaction is determined by PYTHIA, i.e. the direct and resolved photon-nucleon interactions lead to the excitation of one or more hadronic strings which fragment according to the Lund fragmentation scheme[13]. Each time a quark-antiquark pair is produced in the fragmentation process a string splits into two color neutral fragments. If such a color neutral fragment has the mass of a hadron we call it a prehadron. After the formation time the hadronic wave-function has build up and the prehadron has turned into a hadron with the same energy and momentum. In Ref. 14 we have developed a method to extract the complete space-time information of each string fragmentation in PYTHIA. Although our results should not be overstressed since one applies a semi-classical picture to a quantum mechanical problem, our approach makes it

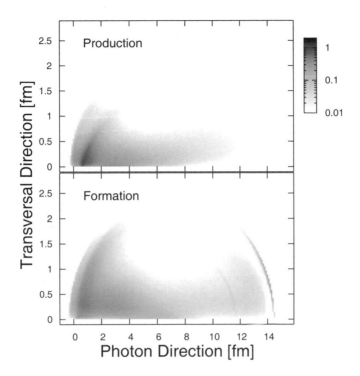

Figure 1. Production (top) and formation (bottom) points for a typical Hermes event: (photon energy $\nu = 14$ GeV, virtuality $Q^2 = 2.5$ GeV^2). The target nucleon is located at the origin, the virtual photon is coming in from the left.

possible to assign the four-dimensional production and formation point to each single hadron in each single scattering event. Figure 1 shows the spatial distribution of the prehadron production and hadron formation points for a typical HERMES event. Obviously, a large fraction of these points fall within a distance behind the struck nucleon that is comparable to nuclear radii.

Starting from their production point the prehadrons propagate through the nucleus and interact with the surrounding nuclear matter. In this work we set their interaction cross section to the corresponding hadronic value. The propagation is described within our semi-classical BUU transport model[9,15] which allows for a probabilistic coupled-channel description of the FSI. Each time a particle interacts with a bound nucleon it might either scatter elastically or produce new particles that are also propagated. At high energies the final state of an inelastic collision is determined by the event generator FRITIOF which is also based on the Lund fragmentation scheme. Finally, we end up with a complete lepton-nucleus event that can be corrected for experimental cuts and detector acceptance.

3. Results

The observable of interest is the so-called multiplicity ratio

$$R_M^h(z_h, \nu) = \frac{\frac{N_h(z_h, \nu)}{N_e(\nu)}\Big|_A}{\frac{N_h(z_h, \nu)}{N_e(\nu)}\Big|_D}, \tag{1}$$

where N_h is the yield of semi-inclusive hadrons in a given (z_h, ν)-bin and N_e the yield of inclusive deep inelastic scattering leptons in the same ν-bin. The quantity $z_h = E_h/\nu$ denotes the energy fraction of the hadron. For the deuterium target, i.e. the nominator of Eq. (1), we simply use the isospin averaged results of a proton and a neutron target. Thus in the case of deuterium we neglect the FSI of the produced hadrons and also the effect of shadowing and Fermi motion.

Figure 2 shows the charged hadron multiplicity ratio for a 27.6 GeV positron beam incident on a ^{14}N and ^{84}Kr target. In our calculations we assume the same kinematic cuts on the scattered lepton and produced hadrons as in the HERMES experiment. In addition we take into account the geometrical acceptance of the HERMES detector. The solid curve shows the result using the default PYTHIA parameter set. In the calculation represented by the dashed curve we used the PYTHIA parameters that have been fitted to hydrogen data by the HERMES collaboration[16]. Both calculations yield comparable results that are in good agreement with the experimental data.

4. Conclusions

Our results demonstrate that one needs large prehadronic cross sections to describe the experimentally observed hadron attenuation in nuclear DIS at HERMES when using the prehadron production points from PYTHIA. In this work we have simply set the prehadronic cross sections to the corresponding hadronic values. So far we have neglected all FSI of the string and string fragments prior to the production of the prehadrons and assumed that string fragmentation in nuclei does not differ from that in vacuum. Both effects might increase the hadron attenuation[7]. In future work one might also think of incorporating induced gluon radiation prior to the

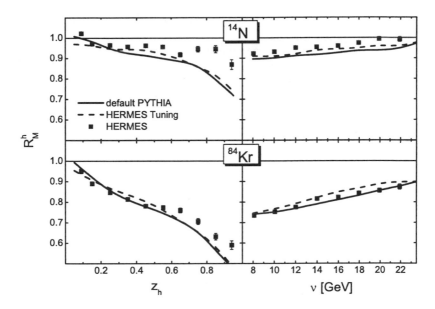

Figure 2. Calculated multiplicity ratio of charged hadrons for ^{14}N and ^{84}Kr nuclei using the four-dimensional production points from the PYTHIA model and setting the prehadronic cross section to the hadronic value. The solid line shows the result using the default PYTHIA parameters. For the calculation represented by the dashed line the fitted parameter set of Ref. 16 has been used. The data are taken from Ref. 3.

production of the color neutral prehadrons. However, one should prevent double counting when considering both gluon bremsstrahlung and the FSI of the hadronic string.

Acknowledgments

Work supported by BMBF. T.F. is supported by the Alexander von Humboldt Foundation (Feodor Lynen Research Fellowship).

References

1. B. Kopeliovich, J. Nemchik, and E. Predazzi, in *Proceedings of the workshop on Future Physics at HERA*, edited by G. Ingelman, A. De Roeck, R. Klanner, DESY, 1995/96, vol 2, p. 1038, nucl-th/9607036.
2. K. Gallmeister, C. Greiner, Z. Xu, Phys. Rev. C67 (2003) 044905; W. Cassing, K. Gallmeister, and C. Greiner, Nucl. Phys. A **735**, 277 (2004); J. Phys. G **30**, S801 (2004).

126

3. A. Airapetian et al. [HERMES Collaboration], Eur. Phys. J. C **20**, 479 (2001); Phys. Lett. B **577**, 37 (2003).
4. W. K. Brooks, Fizika B **13**, 321 (2004).
5. A. Deshpande, R. Milner, R. Venugopalan and W. Vogelsang, hep-ph/0506148.
6. E. Wang, X.-N. Wang, Phys. Rev. Lett. **89**, 162301 (2002); F. Arleo, Eur. Phys. J. C **30**, 213 (2003).
7. A. Accardi, V. Muccifora and H.-J. Pirner, Nucl. Phys. A **720**, 131 (2003); B. Z. Kopeliovich, J. Nemchik, E. Predazzi and A. Hayashigaki, Nucl. Phys. A **740**, 211 (2004); A. Accardi, D. Grunewald, V. Muccifora and H. J. Pirner, hep-ph/0502072.
8. T. Falter, W. Cassing, K. Gallmeister and U. Mosel, Phys. Lett. B **594**, 61 (2004).
9. T. Falter, W. Cassing, K. Gallmeister and U. Mosel, Phys. Rev. C **70**, 054609 (2004).
10. T. Sjöstrand, P. Eden, C. Friberg, L. Lönnblad, G. Miu, S. Mrenna, E. Norrbin, Comp. Phys. Commun. 135 (2001) 238; T. Sjöstrand, L. Lönnblad and S. Mrenna, LU TP 01-21 [hep-ph/0108264].
11. Hong Pi, Comput. Phys. Commun. **71**, 173 (1992); B. Andersson, G. Gustafson, and Hong Pi, Z. Phys. C **57**, 485 (1993).
12. M. Effenberger and U. Mosel, Phys. Rev. C **62**, 014605 (2000); T. Falter and U. Mosel, Phys. Rev. C **66**, 024608 (2002); T. Falter, K. Gallmeister, and U. Mosel, Phys. Rev. C **67**, 054606 (2003).
13. B. Andersson, G. Gustafson, G. Ingelman, T. Sjöstrand, Phys. Rept. 97 (1983) 31; B. Anderson, in: *The Lund Model*, Cambridge University Press (1998).
14. K. Gallmeister and T. Falter, Phys. Lett. B (in press), nucl-th/0502015.
15. M. Effenberger, E. L. Bratkovskaya, and U. Mosel, Phys. Rev. C **60**, 44614 (1999).
16. Patricia Liebing, Ph.D. Thesis, 2004, University of Hamburg.

QUARK GLUON PLASMA AND HADRON GAS ON THE LATTICE

M.P. LOMBARDO

Istituto Nazionale di Fisica Nucleare
Laboratori Nazionali di Frascati
Via Enrico Fermi 40, I-00044, Frascati(RM) Italy
E-mail: lombardo@lnf.infn.it

An informal overview geared towards experimental colleagues, focussed on ab initio calculations and results on quark gluon plasma and the hadronic phase at high temperature

1. Introduction

Up to which extent the properties of the matter produced in the relativistic heavy ion collisions [1] can be predicted by the basic theory of strong interactions, Quantum Chromo Dynamics? In this note we address this point with emphasis on basic idea and results, and we refer the reader to recent reviews [2] for more complete and technical presentations.

QuantumChromoDynamics (QCD) basic degrees of freedom are quarks and gluons. The Lagrangian built by use of these fundamental fields enjoys local and (approximate) global symmetries. The realisation of the global chiral symmetry depends on the thermodynamic conditions of the system : it is spontaneously broken, with the accompanying phenomena of a Goldstone mode and a mass gap, in ordinary conditions, and it gets restored at high temperature. At the same time, confinement, which is realised in the normal phase, disappears at high temperature: all in all, in our low temperature world quarks are confined within hadrons, there is one light preudoscalar meson, the Goldstone boson, the pion, and there are massive mesons and baryons. In the high temperature phase - the Quark Gluon Plasma - quarks and gluons are no longer confined, and the mass spectrum reflects the symmetries of the Lagrangian.

In a standard nuclear physics approach these features of the two phases are imposed by fiat, and the phase transition is obtained by equating the free energies of the quark-gluon gas on one side, and the hadron gas on the other side of the transition. At a variance with this, an approach based on QCD derives the different degrees of freedom of the two phases, as well as the phase transition line, from the same Lagrangian. These calculations, being completely non–perturbative, require a specific technique, Lattice QCD.

Without entering into the details of this approach [3], let me just remind that the QCD equations are put on a 'grid' which should be fine enough to resolve details, and large enough to accommodate hadrons within: obviously, this would call for grids with a large number of points. On the other hand, the calculations complexity grows fast with the number of nodes in the grid, and the actual choices rely on a compromise between physics requirements and computer capabilities.

In practical numerical approaches, this discretisation is combined with a statistical techniques for the computation of the physical observables. This requires a positive 'measure', which is given by the exponential of the Action. A notorious problem plagues these calculations at finite baryon density, as the Action itself becomes complex, with a non–positive definite real part: for many years QCD at nonzero baryon density was not progressing at all. Luckily, in the last four years a few lattice techniques – imaginary chemical potential, Taylor expansion, multiparameter reweighting – proven successful for $\mu_B/T < 1$. [4,5,6,7,8,9,10]. It has to be stressed, however, that these techniques are just dodges and workaround, and do not provide a real solution. Moreover, due to the computer limitations sketched above, which we hope will be soon overcome by the next generation of supercomputers [11], the results have not yet reached the continuum, infinite volume limit.

While waiting for final results in the scaling limit and with physical values of the parameters, it is very useful to contrast and compare current lattice results with model calculations and perturbative studies. The imaginary chemical potential approach[12,13,8,4,5,14]to QCD thermodynamics seems to be ideally suited for the interpretation and comparison with analytic results. Results from an imaginary μ have been obtained for the critical line of the two, three and two plus one flavor model [8], as well as for four flavor [4]. Thermodynamics results – order parameter, pressure, number density – were obtained for the four flavor model [5].

2. *The Hot Phase and the approach to a Quark Gluon Plasma*

The behaviour of the number density at high temperature approaches the lattice Stephan-Boltzmann prediction, with some residual deviation. The deviation from a free field behavior can be parametrized as [17,18]

$$\Delta P(T,\mu) = f(T,\mu)P^L_{free}(T,\mu) \tag{1}$$

where $P^L_{free}(T,\mu)$ is the lattice free result for the pressure. For instance, in the discussion of Ref. [18]

$$f(T,\mu) = 2(1 - 2\alpha_s/\pi) \tag{2}$$

and the crucial point was that α_s is μ dependent.

With the imaginary chemical potential approachs, we can search for such a non trivial prefactor $f(T,\mu)$ by taking the ratio between the numerical data and the lattice free field result $n^L_{free}(\mu_I)$ at imaginary chemical potential:

$$R(T,\mu_I) = \frac{n(T,\mu_I)}{n^L_{free}(\mu_I)} \tag{3}$$

A non-trivial (i.e. not a constant) $R(T,\mu_I)$ would indicate a non-trivial $f(T,\mu)$.

The results for $T \geq 1.5T_c$ seem consistent with a free lattice gas, with an fixed effective number of flavors $N^{eff}_f(T)/4 = R(T)$: $N^{eff}_f = 0.92 \times 4$ for $T = 3.5T_c$, and $N^{eff}_f = 0.89 \times 4$ for $T = 1.5T_c$.

Similar conclusions can be reached via an expanded reweighting technique [6] and via a Taylor expansion [9,23].

2.1. $T_c < T < 1.5T_c$: *a Strongly Coupled Quark Gluon Plasma?*

Of particular interest is the region comprised between T_c and $1.5T_c$: there, it has been noted since long the persistence of of bound states. This does not contradict deconfinement, as heavy states are mostly bound by the short range, Coulombic componenet of the potential. More recently, it has been proposed that coloured states might form as well in this region, further enriching the picture of this state, which has been dubbed strong interactive Quark Gluon Plasma [22]. Lattice studies based on the Taylor expansion [23] do not seem to support this picture, while calculations at imaginary chemical potential are in progress [25]

3. The Hot Phase and the Hadron Resonance Gas Model

At high temperature, but still in the hadronic phase, the system might be described by a gas of weakly interacting resonances, the Hadron Resonance Gas (HRG). The grand canonical partition function of the HRG model[16] has a simple hyperbolic cosine behaviour

$$\Delta P \propto (\cosh(\mu_B/T) - 1) \qquad (4)$$

This behaviour can be assessed eiher via a Taylor expansion of the observables and via imaginary chemical potential calculations. The discussion of the phase diagram in the temperature-imaginary chemical potential plane suggests indeed to use Fourier analysis in this region, as observables are periodic and continuous there[4].

The Fourier analysis of the chiral condensate [4] and of the number density[5] - shows that one cosine [sine] fit describes reasonably well the data up to $T \simeq 0.985 T_c$. This means that the pressure of the model is well approximated by the hadron resonance gas prediction in the broken phase up to $T \simeq 0.985 T_c$.

Analogous results have been obtained by use of a Taylor expansion [21].

3.1. *The Critical Line from HRG*

Kogut and Toublan [24] use the hadron resonance gas model with a fixed energy criterium to draw the phase diagram : in practice, the critical line is implicitly defined by

$$\epsilon(T, \mu) = \epsilon_c \qquad (5)$$

where $\epsilon(T, \mu)$ is the internal energy, and its critical value can be computed on the lattice [4,5]. by analytically continuing the Fourier fits described above: for observables which are even (O_e) or odd (O_o) under $\mu \to -\mu$ the analytic continuation to real chemical potential of the Fourier series read $O_e[o](\mu_I, N_t) = \sum_n a_F^{(n)} \cosh[\sinh](n N_t N_c \mu_I)$. The analytic continuation of any observable O is valid within the analyticity domain, i.e. till $\mu < \mu_c(T)$, where $\mu_c(T)$ has to be measured independently. The value of the analytic continuation of O at μ_c, $O(\mu_c)$, defines its critical value[4,5]. Only calculations at realistic values of the quark masses, close to the continumm limit, will confirm or disprove the fixed energy criterium used in these calculations.

4. Summary

Let me start this summary with a cautionary remark: lattice simulations still need being tuned towards the physical values of the quark masses, and the continuum limit.

This said, the main results obtained so far can be summarized as follows:

- The slope of the critical line in the T, μ plane has been nicely computed [2]. Although this discussion was not in the scope of this talk where I have emphasized the quark–hadron contents of the different phases, it is worthwhile to mention this result. The precise location of the expected endpoint still has to settle down. Results are continuing evolving [26].

- **Revenge of Nuclear Physics !** In the introduction I have emphasized how the Lattice approach makes uses of fundamental degrees of freedom, as opposed to the nuclear physics approach where the relevant degrees of freedom and their interactions follow from a phenomenological analysis: however, it turns out that the hadronic phase is well described by a simple gas of resonances up to $T \simeq 0.98 T_c$!

- For $T \geq 1.5 T_c$ the results are compatible with lattice Stefan Boltzmann with an effective fixed number of flavor slightly below the physical one: these finite density corrections appear to leave unchanged the free field structure.

- The region right above T_c is richer than expected: it might well be the domain of the *Strongly Interactive Quark Gluon Plasma*, characterised by highly nonperturbative features of termodynamics, and by the persistence of bound states, possibly including colored ones.

Finally, we should remind ourselves that the lattice approach to QCD phases and phase transitions discussed in this note is so far limited to equilibrium calculations. One main challenge ahead of us is then to link the static properties measured on the Lattice with the real time evolution during Ultrerelativistic Heavy Ion Collisions [1].

Acknowledgments

It is a pleasure to thank the Organisers for a very nice and interesting meeting.

132

References

1. P. Giubellino, this Volume.
2. see F. Wilczek, Proc. Nat. Acad. Sci. **102** (2005) 8403 [arXiv:hep-ph/0502113] for a general, yet concise introduction; E. Laermann and O. Philipsen, Ann. Rev. Nucl. Part. Sci. **53** (2003) 163 for an excellent review and discussions on the various lattice methods; J. B. Kogut and M. A. Stephanov, Camb. Monogr. Part. Phys. Nucl. Phys. Cosmol. **21** (2004) 1 for an exhaustive discussions on the phases of QCD, with emphasys on the lattice approach; The proceedings of the Conference series on QuarkMatter and Lattice annual Symposium offer extensive reviews on recent results.
3. see e.g. J. Smit, Cambridge Lect. Notes Phys. **15** (2002) 1.
4. M. D'Elia and M. P. Lombardo, Phys. Rev. D **67** (2003) 014505.
5. M. D'Elia and M. P. Lombardo, Phys. Rev. D **70** (2004) 074509.
6. Z. Fodor and S. D. Katz, Phys. Lett. B **534** (2002) 84; JHEP **0404** (2004) 50;
   Z. Fodor, S. D. Katz and K. K. Szabo, Phys. Lett. B **568** (2003) 73;
   F. Csikor *et al.* JHEP **0405** (2004) 046;
   S. D. Katz, Nucl. Phys. Proc. Suppl. **129** (2004) 60.
7. Ph. de Forcrand *et al.*, Nucl. Phys. Proc. Suppl. **119** (2003) 541.
8. Ph. de Forcrand and O. Philipsen, Nucl. Phys. B **642** (2002) 290;
   Nucl. Phys. B **673** (2003) 170.
9. C. R. Allton *et al.*, Phys. Rev. D **66** (2002) 067801;
   Phys. Rev. D **68**, (2003) 014081:
10. R. Gavai, S. Gupta and R. Roy, Prog. Theor. Phys. Suppl.**153** (2004) 270.
11. T. Wettig, talk at Lattice2005, Dublin, to appear in the Proceedings.
12. M. P. Lombardo, Nucl. Phys. Proc. Suppl. **83**, 375 (2000).
13. A. Hart, M. Laine and O. Philipsen, Phys. Lett. B **505**, 141 (2001).
14. P. Giudice and A. Papa, Phys. Rev. D **69**, 094509 (2004).
15. J. B. Kogut et al. Nucl. Phys. B **225**, 93 (1983).
16. F. Karsch, K. Redlich and A. Tawfik, Phys. Lett. B **571**, 67 (2003).
17. K. K. Szabo and A. I. Toth, JHEP **0306**, 008 (2003).
18. J. Letessier and J. Rafelski, Phys. Rev. C **67**, 031902 (2003).
19. A. Vuorinen, arXiv:hep-ph/0402242;
    A. Vuorinen, Phys. Rev. D **68**, 054017 (2003);
    A. Ipp, A. Rebhan and A. Vuorinen, Phys. Rev. D **69**, 077901 (2004).
20. A. Mocsy, F. Sannino and K. Tuominen, Phys. Rev. Lett. **92**, 182302 (2004)
21. C. R. Allton *et al.*, Phys. Rev. D **71**, 054508 (2005)
22. E. Shuryak, J. Phys. G **30** (2004) S1221.
23. S. Ejiri, F. Karsch and K. Redlich, arXiv:hep-ph/0509051; F. Karsch, talks at Lattice 2005, Dublin, July 2005 and QuarkMatter 2005, Budapest, August 2005, to appear in the proceedings.
24. D. Toublan and J. B. Kogut, Phys. Lett. B **605** (2005) 12
25. M. D'Elia, F. Di Renzo and M.P. Lombardo, talk at QCD@Work 2005, Conversano, June 2005 to appear in the Proceedings.
26. O. Philipsen, talk at Lattice2005, Dublin, July 2005, to appear in the Proceedings.

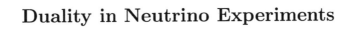

Duality in Neutrino Experiments

NEUTRINOS AND LOCAL DUALITY*

F. M. STEFFENS

NFC - FCBEE - Universidade Presbiteriana Mackenzie,
Rua da Consolação 930,
São Paulo, SP 01302-907, Brazil
E-mail: fsteffen@ift.unesp.br

K. TSUSHIMA

National Center for Theoretical Sciences at Taipei,
Taipei 10617, Taiwan

We report the calculation of the proton and neutron unpolarized structure functions using local duality for both neutrino and muon scattering, in the large x region. Our result indicate a possible violation of local duality at the nucleon pole for the case of neutrino scattering

Our particular interest in the present work is to investigate to what extent local quark-hadron duality can be applied to the proton structure function as measured by neutrinos. To this end, at large x, and assuming charge symmetry at the parton level, we should have in leading order QCD:

$$F_2^{\nu N}(x \to 1) \simeq x[u(x) + \overline{u}(x) + d(x) + \overline{d}(x)], \tag{1}$$

where we did not write the contribution from the strange quarks because they are not expected to contribute in this region [1,2]. On the other hand, the iso-scalar structure function $F_2^{\mu N}$, measured in muon scattering should be given, at large x, by:

$$F_2^{\mu N}(x \to 1) \simeq \frac{5}{18}x[u(x) + \overline{u}(x) + d(x) + \overline{d}(x)], \tag{2}$$

implying

$$\frac{5}{18}F_2^{\nu N}(x \to 1) \simeq F_2^{\mu N}(x \to 1) \tag{3}$$

---

*This work is supported by fapesp (03/10754-0) and cnpq (308932/2003-0)

A failure of this identity would suggest: (1) an unexpected strange distribution at large x; (2) charge symmetry violation in the valence quark distribution; (3) violation of local quark-hadron duality when calculating $F_2^{\nu N}$ and $F_2^{\mu N}$. Any of these three possibilities are very significant and justify a further investigation of the structure functions as probed by neutrinos and muons in the large x region.

As local quark-hadron duality relates the cross section measured in the deep inelastic region to the cross section measured in the resonance region, including the elastic peak, we start writing the hadronic tensor that enters in the quasi elastic neutrino-nucleon cross section, $\nu_\mu(\overline{\nu}_\mu) + n(p) \rightarrow \mu^-(\mu^+) + p(n)$. Keeping only the relevant and sufficient terms for this study [3], we start with the matrix elements of the charged current, which is given by:

$$
\begin{aligned}
< p(P')|J_+^\mu(0)|n(P) > &= < n(P')|J_-^\mu(0)|p(P) > \\
&= \overline{u}(P') \left[F_1^V(Q^2)\gamma^\mu + \frac{i\sigma^{\mu\nu}q_\nu}{2M} F_2^V(Q^2) \right. \\
&\left. - G_A(Q^2)\gamma^\mu\gamma_5 \right] u(P),
\end{aligned}
\tag{4}
$$

where $F_1^V(Q^2)$ and $F_2^V(Q^2)$ are, respectively, the iso-vector Dirac and Pauli form factors, and $G_A(Q^2)$ the axial form factor. The elastic part of the hadronic tensor calculated from charged current is then:

$$
W_{\mu\nu}^{el} = -F_1^{el}\frac{g_{\mu\nu}}{M} + F_2^{el}\frac{P_\mu P_\nu}{2M^3\tau} + iF_3^{el}\varepsilon_{\mu\nu\alpha\beta}\frac{P^\alpha q^\beta}{4M^3\tau},
\tag{5}
$$

with $\tau = Q^2/4M^2$ and

$$
F_1^{el} = \frac{M}{2}\delta\left(\nu - \frac{Q^2}{2M}\right)[\tau(G_M^V)^2 + (1+\tau)G_A^2],
\tag{6}
$$

$$
F_2^{el} = M\tau\delta\left(\nu - \frac{Q^2}{2M}\right)\left[\frac{(G_E^V)^2 + \tau(G_M^V)^2}{1+\tau} + G_A^2\right],
\tag{7}
$$

$$
F_3^{el} = M\tau\delta\left(\nu - \frac{Q^2}{2M}\right)[2G_M^V G_A],
\tag{8}
$$

where ν in Eqs. (6)-(8) is the energy transfer between the beam and the target. The iso-vector electric and magnetic form factors are given by $G_{E,M}^V = G_{E,M}^p - G_{E,M}^n$, with $G_E^N = F_1^N - \tau F_2^N$ and $G_M^N = F_1^N + F_2^N$.

If local quark-hadron duality is valid for the nucleon pole only we have [4]:

$$F_2^{\nu N}(x_{th}) \simeq -\beta \left[\frac{(G_M^V)^2 - (G_E^V)^2}{4M^2(1+\tau)^2} \right.$$
$$\left. + \frac{1}{1+\tau} \left(\frac{d(G_E^V)^2}{dQ^2} + \tau \frac{d(G_M^V)^2}{dQ^2} \right) + \frac{dG_A^2}{dQ^2} \right], \qquad (9)$$

with $\beta = (Q^4/M^2)(\xi_0^2/\xi_{th}^3)(2 - \xi_{th}/x_{th})/(4 - 2\xi_0)$ and $x_{th} = Q^2/(Q^2 + m_\pi(2M + m_\pi))$. A similar calculation can be made for the $F_2^{\mu p}$ structure function [5,4] which combined with Eq. (9) gives:

$$\frac{5}{18} F_2^{\nu N}(x = x_{th} \to 1) - F_2^{\mu N}(x = x_{th} \to 1) \simeq$$
$$+ \frac{13}{18}\beta \left(\frac{d(G_M^p)^2}{dQ^2} + \frac{d(G_M^n)^2}{dQ^2} \right) + \frac{5}{9}\beta \frac{d(G_M^p G_M^n)}{dQ^2} - \frac{5}{18}\beta \frac{dG_A^2}{dQ^2}. \qquad (10)$$

The result of Eq. (10) implies in the violation of Eq. (3). To see how large is this violation, we used a world data parametrization [6] to calculate the form factors. The result, shown in Figure 1, is clearly different from 1, although the calculation should not be trusted for $x = x_{th} \lesssim 0.7$, where $Q^2 \lesssim 0.5\ GeV^2$. However, at $x = x_{th} \sim 0.9$, $Q^2 \sim 2.5\ GeV^2$ and $W^2 \sim 1.25\ GeV^2$, a region where local quark - hadron duality is observed to work within a 10 % accuracy margin [7,8]. The effect shown in Figure 1 is somewhat larger than the known limitations of local quark - hadron duality. If taken seriously, they show an effect at large x that is not marginal. What is the origin of this effect? If it comes from charge symmetry violation, it can be shown that the violation in the u distribution should be larger than the violation in the d distribution [4], although that in order to fit the result in Figure 1, it would have to be too large, in disagreement with others theoretical estimates [9]. More likely is the possibility that local duality for the nucleon pole, in the case of neutrino scattering, does not hold, in opposition to the case of electron scattering [10].

References

1. S. Kretzer, H. L. Lai, F. I. Olness and W. K. Tung, Phys. Rev. D **69**, 114005 (2004); A. D. Martin, R. G. Roberts, W. J. Stirling and R. S. Thorne, Phys. Lett. B **531**, 216 (2002).
2. V. Barone, C. Pascaud and F. Zomer, Eur. Phys. J. C **12**, 243 (2000).
3. A. W. Thomas and W. Weise, *The Structure of the Nucleon*, Berlin-Germany, Wiley & Sons (2001).
4. F. M. Steffens and K. Tsushima, Phys. Rev. **D70**, 094040 (2004).

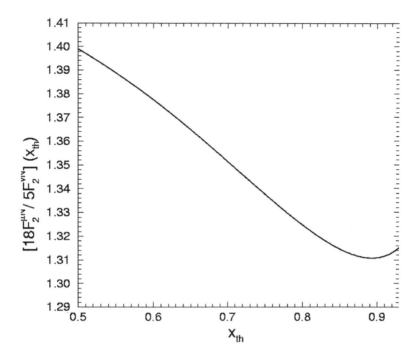

Figure 1. The ratio between the iso-scalar structure function as probed by neutrino and muon beams. In leading order QCD, this ratio should approach 1 as $x = x_{th} \to 1$

5. W. Melnitchouk, Phys. Rev. Lett. **86**, 35 (2001).
6. K. Tsushima, Hungchong Kim and K. Saito, nucl-th/0307013, to appear in Phys. Rev. C; M. J. Musolf and T. W. Donelly, Nucl. Phys. A **546**, 509 (1992); T. Kitagaki et al., Phys. Rev. D **28**, 436 (1983).
7. I. Niculescu et al., Phys. Rev. Lett. **85**, 1186 (2000).
8. N. Bianchi, A. Fantoni and S. Liuti, Phys. Rev. D **69**, 014505 (2004).
9. E. N. Rodionov, A. W. Thomas and J. T. Londergan, Mod. Phys. Lett. A **9**, 1799 (1994); J. T. Londergan, G. T. Carvey, G. Q. Liu, E. N. Rodionov and A. W. Thomas, Phys. Lett. B **340**, 115 (1994).
10. W. Melnitchouk, these proceedings.

Duality and QCD

HIGHER TWIST EFFECTS IN POLARIZED DIS [*]

E. LEADER

Imperial College London
Prince Consort Road, London SW7 2BW, England
E-mail: e.leader@imperial.ac.uk

A. V. SIDOROV

Bogoliubov Theoretical Laboratory
Joint Institute for Nuclear Research, 141980 Dubna, Russia
E-mail:sidorov@thsun1.jinr.ru

D. B. STAMENOV

Institute for Nuclear Research and Nuclear Energy
Bulgarian Academy of Sciences
Blvd. Tsarigradsko Chaussee 72, Sofia 1784, Bulgaria
E-mail: stamenov@inrne.bas.bg

The size of higher twist corrections to the spin proton and neutron g_1 structure functions and their role in determining the polarized parton densities in the nucleon is discussed.

1. Introduction

The study of the quark-hadron duality phenomena[1] using the present much more precise data for the polarized and unpolarized structure functions is in a progress[2]. The better understanding of the non-perturbative higher twist effects is important for this analysis, especially in the polarized case, where the investigations are in the very beginning.

In this talk we will discuss both, the size of the higher twist effects in *polarized* DIS as well as their role in the determination of the polarized parton densities (PPDs) in the nucleon using different approaches of QCD fits to the data.

---

[*]This research was supported by the UK Royal Society and JINR-Bulgaria Collaborative Grants and by the RFBR (no 05-01-00992, 03-02-16816).

2. QCD Treatment of the g_1 Structure Function

In QCD the spin structure function g_1 can be written in the following form ($Q^2 >> \Lambda^2$):

$$g_1(x, Q^2) = g_1(x, Q^2)_{\text{LT}} + g_1(x, Q^2)_{\text{HT}} , \tag{1}$$

where "LT" denotes the leading twist ($\tau = 2$) contribution to g_1, while "HT" denotes the contribution to g_1 arising from QCD operators of higher twist, namely $\tau \geq 3$. In Eq. (1) (the nucleon target label N is dropped)

$$g_1(x, Q^2)_{\text{LT}} = g_1(x, Q^2)_{\text{pQCD}} + h^{\text{TMC}}(x, Q^2)/Q^2 + \mathcal{O}(M^4/Q^4) , \tag{2}$$

where $h^{\text{TMC}}(x, Q^2)$ are the calculable[3] kinematic target mass corrections, which effectively belong to the LT term. $g_1(x, Q^2)_{\text{pQCD}}$ is the well known (logarithmic in Q^2) pQCD expression and in NLO has the form

$$g_1(x, Q^2)_{\text{pQCD}} = \frac{1}{2} \sum_q^{N_f} e_q^2[(\Delta q + \Delta \bar{q}) \otimes (1 + \frac{\alpha_s(Q^2)}{2\pi} \delta C_q) + \frac{\alpha_s(Q^2)}{2\pi} \Delta G \otimes \frac{\delta C_G}{N_f}], \tag{3}$$

where $\Delta q(x, Q^2), \Delta \bar{q}(x, Q^2)$ and $\Delta G(x, Q^2)$ are quark, anti-quark and gluon polarized densities in the proton, which evolve in Q^2 according to the spin-dependent NLO DGLAP equations. $\delta C(x)_{q,G}$ are the NLO spin-dependent Wilson coefficient functions and the symbol $\otimes$ denotes the usual convolution in Bjorken x space. N_f is the number of active flavors.

In Eq. (1)

$$g_1(x, Q^2)_{\text{HT}} = h(x, Q^2)/Q^2 + \mathcal{O}(1/Q^4) , \tag{4}$$

where $h(x, Q^2)$ are the *dynamical* higher twist ($\tau = 3$ and $\tau = 4$) corrections to g_1, which are related to multi-parton correlations in the nucleon. The latter are non-perturbative effects and cannot be calculated without using models. That is why a *model independent* extraction of the dynamical higher twists $h(x, Q^2)$ from the experimental data is important not only for a better determination of the polarized parton densities but also because it would lead to interesting tests of the non-perturbative QCD regime and, in particular, of the quark-hadron duality.

One of the features of polarized DIS is that a lot of the present data are in the preasymptotic region ($Q^2 \sim 1 - 5 \ GeV^2, 4 \ GeV^2 < W^2 < 10 \ GeV^2$). While in the unpolarized case we can cut the low Q^2 and W^2 data in order to minimize the less known higher twist effects, it is impossible to perform such a procedure for the present data on the spin-dependent structure functions without losing too much information. This is especially the case for the HERMES, SLAC and Jefferson Lab experiments. So, to confront correctly the QCD predictions with the experimental data and to determine the *polarized* parton densities special attention must be paid to the non-perturbative higher twist (powers in $1/Q^2$) corrections to the nucleon structure functions.

3. Higher Twists and Their Role in Determining PPDs

We have used two approaches to extract the polarized parton densities from the world polarized DIS data. According to the first[4] the leading twist LO/NLO QCD expressions for the structure functions g_1 and F_1 have been used in order to confront the data on spin asymmetry $A_1(\approx g_1/F_1)$ and g_1/F_1. We have shown[5,6] that in this case the extracted from the world data 'effective' HT corrections $h^{g_1/F_1}(x)$ to the ratio g_1/F_1

$$\left[\frac{g_1(x,Q^2)}{F_1(x,Q^2)}\right]_{exp} \Leftrightarrow \frac{g_1(x,Q^2)_{LT}}{F_1(x,Q^2)_{LT}} + \frac{h^{g_1/F_1}(x)}{Q^2} \tag{5}$$

are negligible and consistent with zero within the errors, *i.e.* $h^{g_1/F_1}(x) \approx 0$, when for $(g_1)_{LT}$ and $(F_1)_{LT}$ their NLO QCD approximations are used. (Note that in QCD the unpolarized structure function F_1 takes the same form as g_1 in (1), namely $F_1 = (F_1)_{LT} + (F_1)_{HT}$.) What follows from this result is that the higher twist corrections to g_1 and F_1 approximately *compensate* each other in the ratio g_1/F_1 and the NLO PPDs extracted this way are less sensitive to higher twist effects. This is not true in the LO case (see our discussion in Ref. 7). The set of polarized parton densities extracted this way is referred to as $PD(g_1^{NLO}/F_1^{NLO})$.

According to the second approach[7], the g_1/F_1 and A_1 data have been fitted using phenomenological parametrizations of the experimental data for the unpolarized structure function $F_2(x,Q^2)$ and the ratio $R(x,Q^2)$ of the longitudinal to transverse γN cross-sections (i.e. F_1 is replaced by its expression in terms of usually extracted from unpolarized DIS experiments F_2 and R). Note that such a procedure is equivalent to a fit to $(g_1)_{exp}$, but it is more consistent than the fit to the g_1 data themselves actually presented by the experimental groups because here the g_1 data are extracted in the same way for all of the data sets. In this case the HT corrections to g_1 cannot be compensated because the HT corrections to $F_1(F_2$ and $R)$ are absorbed in the phenomenological parametrizations of the data on F_2 and R. Therefore, to extract correctly the polarized parton densities from the g_1 data, the HT corrections (4) to g_1 have to be taken into account. So, according to this approach we have used the following expression for the ratio g_1/F_1:

$$\left[\frac{g_1^N(x,Q^2)}{F_1^N(x,Q^2)}\right]_{exp} \Leftrightarrow \frac{g_1^N(x,Q^2)_{LT} + h^N(x)/Q^2}{F_1^N(x,Q^2)_{exp}} , \tag{6}$$

where $g_1^N(x,Q^2)_{LT}$ (N=p, n, d) is given by the leading twist expression (3) in LO/NLO approximation including the target mass corrections. In (6) $h^N(x)$ are the dynamical $\tau = 3$ and $\tau = 4$ HT corrections which are extracted in a *model independent way*. In our analysis their Q^2 dependence is neglected. It is small and the accuracy of the present data does not allow to determine it. The set of PPDs extracted according to this approach is referred to as PD(g_1^{LT} + HT). The details of our recent analysis using the present available data on polarized DIS are given in [8].

The extracted higher twist corrections to the proton and neutron spin structure functions, $h^p(x)$ and $h^n(x)$, are shown in Fig. 1. As seen from Fig. 1 the size of the HT corrections is not negligible and their shape depends

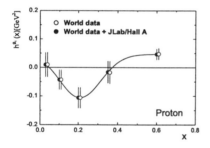 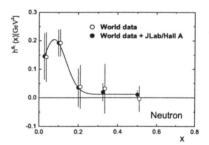

Figure 1. Higher twist corrections to the proton and neutron g_1 structure functions extracted from the data on g_1 in NLO($\overline{\text{MS}}$) QCD approximation for $g_1(x,Q^2)_{LT}$. The parametrization (7) of the higher twist values is also shown.

on the target. In Fig. 1 our previous results on the higher twist corrections to g_1 (before the JLab Hall A data were available) are also presented. As seen from Fig. 1, thanks to the very precise JLab Hall A data[9] at large x the higher twist corrections to the neutron spin structure function are now much better determined in this region. In Fig. 1 our parametrizations of the values of higher twists for the proton and neutron targets

$$h^p(x) = 0.0465 - \frac{0.1913}{\sqrt{\pi/2}} exp[-2((x - 0.2087)/0.2122)^2]$$

$$h^n(x) = 0.0119 + \frac{0.2420}{\sqrt{\pi/2}} exp[-2((x - 0.0783)/0.1186)^2] \qquad (7)$$

are also shown. These should be helpful in a calculation of the nucleon structure function g_1 for any x and moderate Q^2 in the experimental region, where the higher twist corrections are not negligible. The impact of the very recent COMPASS data[13] on the values of higher twist corrections is negligible. The new values are in a good agreement with the old ones[8].

The values of the higher twist corrections to the proton and neutron g_1 structure functions extracted in a model independent way from polarized DIS data are in agreement with the QCD sum rule estimates[10] as well as with the instanton model predictions[11] but disagree with the renormalon calculations[12]. About the size of the HT corrections extracted from the resonance region see the discussion in the Fantoni's talk at this Workshop.

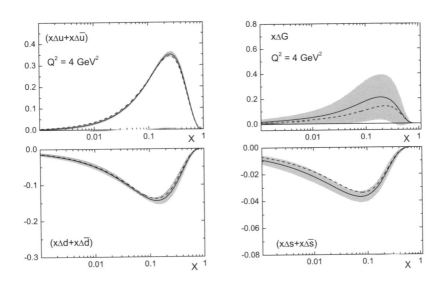

Figure 2. NLO($\overline{\text{MS}}$) polarized parton densities PD(g_1^{LT} + HT) (solid curves) together with their error bands compared to PD($g_1^{\text{NLO}}/F_1^{\text{NLO}}$) (dashed curves) at $Q^2 = 4\ GeV^2$.

In Fig. 2 we compare the NLO($\overline{\text{MS}}$) polarized parton densities PD(g_1^{LT} +

HT) with $PD(g_1^{NLO}/F_1^{NLO})$. As seen from Fig. 2 the two sets of PPDs are very close to each other, especially for u and d quarks. This is a good illustration of the fact that a fit to the g_1 data taking into account the higher twist corrections to g_1 ($\chi^2_{DF,NLO} = 0.872$) is equivalent to a fit of the data on $A_1(\sim g_1/F_1$) and g_1/F_1 using for the g_1 and F_1 structure functions their NLO leading twist expressions ($\chi^2_{DF,NLO} = 0.874$). In other words, this fact confirms once more that the HT corrections to g_1 and F_1 approximately cancel in the ratio g_1/F_1. Nevertheless, we consider that the set of the polarized parton densities $PD(g_1^{LT} + HT)$ is preferable because using them and simultaneously extracted higher twist corrections to g_1, the spin structure function g_1 can be correctly calculated in the preasymptotic (Q^2, W^2) region too.

In conclusion, the higher twist effects in polarized DIS have been studied. It was shown that the size of the HT corrections to the spin structure function g_1 is not negligible and their shape depends on the target. It was also demonstrated that their role is important for the correct determination of the polarized parton densities.

References

1. E.D. Bloom and F.J. Gilman, *Phys. Rev. Lett.* **25**, 1140 (1970); *Phys. Rev.* **D4**, 2901 (1971).
2. A. Fantoni, N. Bianchi and S. Liuti, *AIP Conf. Proc.* **747**, 126 (2005)[hep-ph/0501180]; see also the talk by A. Fantoni at this Workshop.
3. A. Piccione and G. Ridolfi, *Nucl. Phys.* **B513**, 301 (1998); J. Blumlein and A. Tkabladze, *Nucl. Phys.* **B533**, 427 (1999).
4. E. Leader, A.V. Sidorov and D.B. Stamenov, *Eur. Phys. J.* **C 23**, 479 (2002).
5. E. Leader, A.V. Sidorov and D.B. Stamenov, in *Particle Physics at the Start of the New Millennium*, edited by A.I. Studenikin, World Scientific, Singapore, May 2001, p. 76. (*Proceedings of the 9th Lomonosov Conference on Elementary Particle Physics, Moscow, Russia, 20-26 Sep 1999*).
6. E. Leader, A. V. Sidorov and D. B. Stamenov, in *Deep Inelastic Scattering DIS2003*, edited by V.Kim and L.Lipatov, PNPI RAS, 2003, pp. 790-794.
7. E. Leader, A.V. Sidorov and D.B. Stamenov, *Phys. Rev.* **D 67**, 074017 (2003).
8. E. Leader, A.V. Sidorov and D.B. Stamenov, *JHEP* **06**, 033 (2005).
9. JLab/Hall A Coll., X. Zheng *et al.*, *Phys. Rev. Lett.* **92**, 012004 (2004).
10. I.I. Balitsky, V.M. Braun and A.V. Kolesnichenko, *Phys. Lett.* **B242**, 245 (1990), *ibid* **B318**, 648 (1993); E. Stein *et al.*, *Phys. Lett.* **B353**, 107 (1995).
11. J. Balla, M.V. Polyakov and C. Weiss, *Nucl. Phys.* **B510**, 327 (1998); A.V. Sidorov and C. Weiss, hep-ph/0410253.
12. E. Stein, *Nucl. Phys. Proc. Suppl.* **79**, 567 (1999).
13. COMPASS Coll., E.S. Ageev *et al.*, *Phys. Lett.* **B612**, 154 (2005).

HIGHER TWIST EFFECTS IN POLARIZED EXPERIMENTS

NILANGA LIYANAGE

University of Virginia,
Charlottesville, VA 22904, USA
E-mail: nilanga@virginia.edu

Higher twist effects in nucleon spin structure functions provide an opportunity to understand quark-quark and quark-gluon correlations in the nucleon. A series of experiments in Jefferson Lab Hall A has enabled the extraction of higher twist effects in the neutron with unprecedented precision

A series of recent polarized electron scattering experiments performed at Jefferson Lab has provided an extensive data set for spin structure functions g_1, g_2 and virtual photon asymmetries A_1 and A_2 at low and moderate momentum transfers. These data allow for a precision study of the higher twist effects in polarized structure functions.

The higher-twist effects in structure functions can be best understood within the formalism of the Operator Product Expansion (OPE). In this framework g_2 can be separated into a twist-2 and a higher-twist term:

$$g_2(x, Q^2) = g_2^{WW}(x, Q^2) + g_2^{H.T.}(x, Q^2) , \qquad (1)$$

where the leading-twist (twist-2) term, $g_2^{WW}(x, Q^2)$, can be determined using g_1 as[4]

$$g_2^{WW}(x, Q^2) = -g_1(x, Q^2) + \int_x^1 \frac{g_1(y, Q^2)}{y} dy . \qquad (2)$$

The higher-twist term arises from the quark-gluon correlations. Thus g_2 provides a clean way to study higher-twist effects. Furthermore, at high Q^2, the twist-3 matrix element d_2 is derived from the x^2-weighted moment of the twist-3 part of g_2:

$$d_2 = \int_0^1 x^2 [g_2(x) - g_2^{WW}(x)] dx. \qquad (3)$$

The d_2 matrix element is related to the color polarizabilities[5] of the nucleon. Predictions for d_2 exist from various models and lattice QCD.

In this paper we present results for g_2^n, A_2^n and d_2^n from a series of experiments carried out in Jefferson lab Hall A over the last few years. These experiments used the Hall A polarized ^{3}He target as an effective neutron target. The design of this target allows for easy transition into the transverse target polarization mode from the longitudinal polarization mode. Parallel and perpendicular asymmetries were extracted for ^{3}He in this series of experiments. The combination of parallel and perpendicular asymmetries allows for a clean separation of structure functions g_1 and g_2 (and asymmetries A_1 and A_2)[6]. Both Hall A High Resolution spectrometers (HRS) were used in a symmetric configuration in electron detection mode to measure the inclusive $^3\vec{H}e(e,e')X$ reaction.

The measured parallel and perpendicular asymmetries were combined with the beam and target polarization measurements and the dilution factor to extract structure functions ,and in some cases, virtual photon asymmetries for ^{3}He. The most recent models[7] were used to apply nuclear corrections to the data in the DIS region to obtain the corresponding quantities for the neutron. In the case of low Q^2 resonance region data, the nuclear correction were applied to extract moments of the neutron spin structure functions.

Jefferson lab Hall A experiment E97-103[8] made a precision measurement of g_2^n in the DIS region at x $\approx$ 0.2 covering five Q^2 values from 0.58 to 1.36 GeV^2. Results for g_2^n and g_1^n from this experiment are given in Fig. 1. The light-shaded areas in the two plots give the leading-twist contribution to these two quantities. The dark-shaded areas (next to the horizontal axes) show the systematic errors.

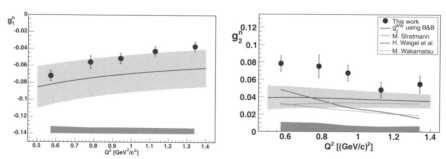

Figure 1.   Results for g_1^n (left) and g_2^n (right) from E97103.

The precision reached in this experiment is more than an order of magnitude improvement over that of previous world data. The deviation of g_2 from the leading twist part (g_2^{WW}) calculations[4] is a measure of higher twist effects arising from quark-gluon correlations. The g_2^{WW} values were obtained from a fit[9] to the high Q^2 world data on g_1^n. This fit was then evolved down to the Q^2 values of this experiment. The measured g_2^n values are consistently higher than g_2^{WW}. For the first time, there is a clear indication of higher-twist effects at the level of precision of these data. The new g_1^n data agree with the leading-twist calculations within the uncertainties.

In 2001, Hall A experiment E99-117[10] made a precision measurement of A_1^n in the high x region from $x = 0.33$ to $x = 0.61$ (Q^2 from 2.7 to 4.8 GeV^2). This data also allowed the extraction of A_2^n. The precision of the A_2^n thus extracted is comparable to that of the best existing world data[11] at high x. Combining these results with the world data, d_2^n was extracted at an average Q^2 of 5 GeV^2:

$$d_2^n = 0.0062 \pm 0.0028. \tag{4}$$

Compared to the previously published result[11], the uncertainty on d_2^n has been improved by about a factor of 2.

The d_2^n measurements also provide the means to study the transition of nucleon spin structure from high Q^2 to low Q^2. At low Q^2, $\bar{d}_2$ is defined to be the inelastic part of $d_2(Q^2)$. At high Q^2, $\bar{d}_2$ becomes d_2 since the elastic contribution is negligible.

Hall A experiment E94-010[12] measured $\bar{d}_2$ (among several other results) at six values of Q^2 from 0.1 to 0.9 GeV^2. In Fig. 2, $\bar{d}_2(Q^2)$ measured in E94-010 and E99-117 are shown along with SLAC results from Experiment E155[11] (solid circles). The grey band represents the systematic uncertainty. The solid line is the MAID[13] calculation containing only the resonance contribution. At low Q^2 a Heavy Baryon χPT calculation[14] (dashed line) is shown. The Lattice QCD prediction[15] at $Q^2 = 5$ GeV^2 is negative but close to zero. At moderate Q^2, our data show that $\bar{d}_2^n$ is positive and decreases with Q^2.

The spin structure data from Jefferson Lab Experimental Halls A and B have been used to calculate twist-4 and twist-6 coefficients of the OPE series of Γ_1, the first moment of g_1, for the proton and the neutron. These coefficients have then been used to extract the twist-4 matrix element f_2. Due to lack of space, the details of this higher twist extraction is not given here and the reader is referred to [20,21,22] and [23]. The results for extracted f_2 for the proton, neutron and $p - n$ are compared to theoretical calculations in Fig. 2 (right).

In summary, polarized experiments in Jefferson Lab Halls A and B have provided precision spin structure data that has enabled an extraction of higher twist effects in the nucleon with unprecedented accuracy. These

150

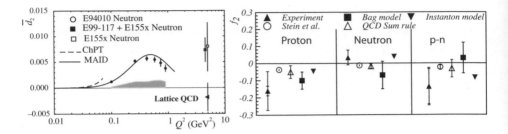

Figure 2. The left panel shows the Hall A results of $\bar{d}_2$ for the neutron along with SLAC data at high Q^2 compared to Lattice QCD, MAID model and HBχPT calculations. The right plot compares f_2 with calculations of Stein et al.[16], MIT bag model[17], QCD sum rules[18] and an instanton model[19].

new results are essential for understanding quark-gluon correlations.

The work presented was supported in part by the U. S. Department of Energy (DOE) contract DE-AC05-84ER40150 Modification NO. M175, under which the Southeastern Universities Research Association operates the Thomas Jefferson National Accelerator Facility.

References

1. N. Isgur, Phys. Rev. D**59**, 034013 (1999).
2. S. Brodsky, M Burkhardt and I. Schmidt, Nucl. Phys. B**441**, 197 (1995).
3. K. Wilson, Phys. Rev. **179**, 1499 (1969).
4. S. Wandzura and F. Wilczek, Phys. Lett. B 72 (1977).
5. X. Ji and P. Unrau, Phys. Lett. B **333**, 228 (1994).
6. The details of the experimental setup can be found in: Phys. Rev. C **70**, 065207 (2004).
7. F. Bissey, et al., Phys. Rev. C **65**, 064317 (2002).
8. K. Kramer, et al., nucl-ex/0506005, submitted to Phys. Rev. Lett.
9. J. Blümlein and H. Bottcher, Nucl. Phys. B **636**, 225 (2002).
10. X. Zheng, et al., Phys. Rev. Lett. **92**, 012004 (2004); Phys. Rev. C **70**, 065207 (2004).
11. K. Abe, et al., E155 collaboration, Phys. Lett. B
12. M. Amarian et al., Phys. Rev. Lett. **89**, 242301 (2002); M. Amarian et al., Phys. Rev. Lett.**92**, 022301 (2004); M. Amarian et al., Phys. Rev. Lett. **93**, 152301 (2004); See also the contribution by Z.-E. Meziani in these proceedings.
13. D. Drechsel, S.S. Kamalov, and L. Tiator, Phys. Rev. D **63**, 114010 (2001).
14. C. W. Kao, T. Spitzenberg and M. Vanderhaeghen, Phys. Rev. D **67**, 016001 (2003).
15. M. Gockeler et al., Phys. Rev. D **63**, 074506, (2001).
16. E. Stein, P. Gornicki, L. Mankiewicz and A. Schäfer, Phys. Lett. B **353**, 107 (1995).
17. X. Ji and W. Melnitchouk, Phys. Rev. D **56**,1 (1997).

18. I.I. Balitsky, V. M. Braun and A.V. Kolesnichenko, Phys. Lett. **B 242**, 245 (1990).
19. N-Y. Lee, K. Goeke and C. Weiss, Phys. Rev. **D 65**, 054008 (2002).
20. Z.-E. Meziani *et al.*, Phys. Lett. **B 613**, 148 (2005).
21. A. Deur, nucl-ex/0508022
22. A. Deur and *et al.*, Phys. Rev. Lett. **93**, 212001 (2004).
23. J. P. Chen, A. Deur, and Z. E. Meziani, nucl-ex/0509007.

STATUS OF POLARIZED AND UNPOLARIZED DEEP INELASTIC SCATTERING*

JOHANNES BLÜMLEIN

Deutsches Elektronen-Synchrotron, DESY, Platanenallee 6, D-15738 Zeuthen, Germany
E-mail: Johannes.Bluemlein@desy.de

The current status of deep inelastic scattering is briefly reviewed. We discuss future theoretical developments desired and measurements needed to further complete our understanding of the picture of nucleons at short distances.

1. Introduction

The discovery of the partonic substructure of nucleons by the SLAC–MIT experiments [1] 35 years ago marks the beginning of the investigation of the nucleon's short distance structure. During the subsequent decades numerous $e^{\pm}N$, $\mu^{\pm}N$ and $\nu(\overline{\nu})N$–experiments were performed at SLAC, FNAL, CERN, DESY and JLAB both for unpolarized and polarized targets to refine our understanding of nucleons in wider and wider kinematic domains and at higher luminosities which allowed rather precise measurements.

Along with this, the theoretical understanding deepened applying Quantum Chromodynamics (QCD) perturbatively to higher orders and investigating some of the related operator matrix elements with non–perturbative methods in the framework of Lattice QCD over the last decades. Deep inelastic scattering data do allow for QCD tests at the 1% level [2] at present, which requires $O(\alpha_s^3)$ accuracy for the perturbative calculations.

In the following I give a brief survey on the present status of deeply inelastic scattering (DIS) and discuss the current challenges for theory and experiment in this field.

---

*Invited talk presented at Duality05, LNFN Frascati, 2005.

154

2. Theory

The theory of deeply inelastic scattering has a history of about 40 years.
Beginning with the early work on the light cone expansion [3] and the parton
model [4] a conclusive picture of the twist–2 contributions [5] arose in complementary languages. With the advent of QCD [6] and finding asymptotic
freedom [7] the scaling violations of nucleon structure functions were studied
systematically.

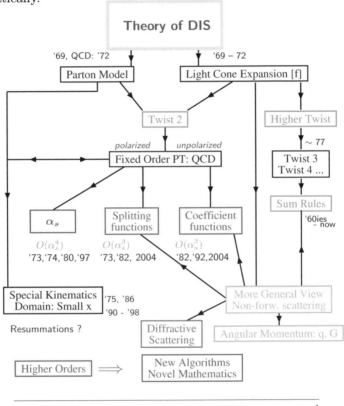

Figure 1: The development of the theory of deeply inelastic scattering. The
years refer to the respective calculations.

The leading order (LO) results for the anomalous dimensions (1973) [8]
were followed by the LO coefficient functions, NLO anomalous dimensions
and coefficient functions [9], see Figure 1, until after about 20 years the 3–
loop anomalous dimensions and coefficient functions could be calculated
recently [10]. These calculations are required to match the current experi-

mental accuracies, in particular to extract the QCD parameter Λ_{QCD} with a theoretical error below the experimental accuracy. Similar timescales were necessary to reach the 4–loop level for the QCD β-function. The step from NLO to NNLO expressions required a significant change in technology and intense use of efficient Computer Algebra programs like FORM [11] due to the proliferation of terms emerging. The calculus of harmonic sums and associated functions [12] was both helpful to design a uniform language for higher order calculations, led to a systematic approach, and gained deeper insight into what eventually is really behind intermediate large expressions generated by Feynman diagrams. Still further progress has to be made in the future. QCD perturbation theory took an enormous development during the last three decades transforming our understanding from an initially qualitative one to highly a quantitative level. Physics, as a quantitative science, knows no other ways but precision calculations to put theoretical ideas and theories to an utmost check. In this way, Quantum Chromodynamics became a well tested, established physical theory, which of course is a process to be continued steadily. The mathematical methods being developed in course to perform this task have a deep beauty and give us insight into quantum field theory on a meta-level. Their uniformal applicability has furthermore led to a quick spread into a series of neighboring fields, as electro–weak theory and string theory, and became, only a few years after their development, a common tool of the community.

With the advent of HERA it became possible to probe the small–x region. Much work was devoted to resummations in this particular region. As pioneered by Lipatov and collaborators [13] both the leading contributions to the splitting and coefficient functions can be derived by arguments of scale invariance. Resummed NLx corrections were calculated in [14]. All these results are of very importance as all–order predictions for the leading and next-to-leading term for splitting functions and Wilson coefficients. The comparison of these predictions with the corresponding results obtained in complete fixed order calculations showed agreement. Furthermore one should stress, that the LO resummations [13] refer to scheme-invariant, i.e. physical quantities, and do thus predict as well corresponding matching conditions between splitting and coefficient functions.

Unfortunately these resummations turn out to be not dominant in the small x region for the description of structure functions, since the respective kernels have to be convoluted with parton densities, which are strongly rising towards small values of x and formally sub-leading terms contribute at the same strengths [15].

For polarized deep–inelastic scattering the anomalous dimensions and coefficient functions are known to NLO at present. Although the statistical and systematic errors for the polarized parton densities is still large, the NNLO improvement would be desirable to further minimize the factorization and renormalization scale uncertainties [16].

Polarized deep inelastic scattering offers access to twist–3 operator matrix elements and predictions for their scale dependence in QCD. At present, the radiative corrections for these terms are worked out in one-loop order. The understanding of QCD higher twist contributions beyond twist 3 both for unpolarized and polarized deep inelastic scattering is still in its infancy and will require more work in the future having more precise data available. Since higher twist anomalous dimensions and coefficient functions refer to more than one ratio of scales but structure functions contain one scale, $x_{\rm Bj}$, only, it is required in general to measure the corresponding operator matrix elements on the lattice at least for the lowest moments.

The light–cone expansion as used in deep inelastic scattering can be generalized to a series of other processes. This more general view concerns non–forward scattering at large space–like virtualities [17]. In this way one may access the angular momentum of partons [18], which is important for the understanding of the spin–structure of nucleons. Moreover, the framework provides several projections on various inclusive quantities of interest. During the last decade a lot of progress has been made in this field calculating the corresponding LO and NLO anomalous dimensions and Wilson coefficients both for the unpolarized and polarized case and understanding conformal symmetry and its breaking for this process in QCD. A related picture was developed also to describe diffractive ep–scattering [19], which yields a proper description of this process using the notion of an observed rapidity gap only, without referring to the concept of a pomeron.

3. Experiment

Deep inelastic scattering has been probed by now in a wide kinematic range : $10^{-5} < x < 0.8$, $4 < Q^2 \lesssim 50.000\,{\rm GeV}^2$. Figure 2 gives an overview on different experiments and facilities showing also the luminosities reached or planned. The proton structure function $F_2^p(x, Q^2)$ is a well measured quantity in all this range. Both to perform flavor separation and QCD tests, it is highly desirable to know the neutron structure function $F_2^n(x, Q^2)$ [20] at comparable accuracy in the same kinematic region. This has been the case for fixed target experiments. Both measurements allow to extract the

u_{val} and d_{val} distributions at comparable precision not only in the valence region $x \gtrsim 0.3$ but also in the region below, supplementing the DIS data with Drell-Yan data on $\bar{d}(x) - \bar{u}(x)$. A non–singlet QCD analysis to $O(\alpha_s^3)$ was performed [21], widely free on assumptions on the gluon and sea-quark densities. The error of α_s is of $\sim 3\%$.

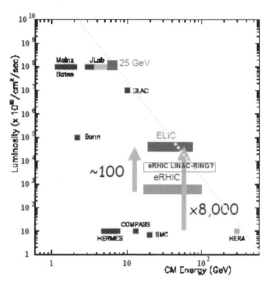

Figure 2: DIS lN scattering experiments: Luminosity vs cms Energy [curtesy R. Ent].

The HERA experiments extended the kinematic region by two orders of magnitude both in x and Q^2. From the high Q^2 large x data the valence distributions will be measured in a yet widely unexplored region, in which potentially new physics may be found. These measurements can be compared to those at lower values of Q^2 by evolution. Statistically very precise measurements were performed in the medium and lower Q^2 and small x region, which gives access to the charge–weighted sea–quark and gluon distributions. As the detailed knowledge of the gluon and sea–quark distributions is instrumental to the future physics at LHC, HERA plays a key role in determining these quantities. In the case of $F_2(x, Q^2)$ the gluon distribution enters indirectly and determines the slope in $\ln(Q^2)$ of the structure function rather its value. From the measurement of both the slope and value of the structure function one may uniquely unfold the gluon and sea quark contribution without reference to a priori choices of shapes, cf. [22]. Further important input for a measurement of the gluon distribution come from precise data on $F_2^{c\bar{c}}(x, Q^2)$ and $F_L(x, Q^2)$. For both these structure functions the gluon distribution enters linearly already in the lowest order. Combined non-singlet and singlet NNLO QCD analyses, partly including collider data, were performed in [23,24] measuring α_s to a precision of $2 - 3\%$ with central values in complete accordance with that of the non–singlet analysis [21].

The singlet quark distribution can well be extracted from ep–structure function measurements. On the other hand, the flavor structure of the sea quark distributions is hard to be resolved in neutral current interactions. Here future high luminosity neutrino experiments at high energy will contribute. Drell–Yan data [25] provide information on the difference $\bar{d}(x) - \bar{u}(x)$. Still higher precision data is needed to resolve as well the Q^2 dependence. Information about the strange quark distributions $s(x)$, $\bar{s}(x)$ is currently gotten in a rather indirect way from the di–muon sample in DIS neutrino scattering. The high statistics measurements stem from iron targets and very little is known on the EMC–effect on strangeness in the lower x region. The charm and beauty quark production in deep inelastic ep scattering is well described by the heavy flavor Wilson coefficients calculated to NLO [26]. Very recently the NNLO corrections in the case of the longitudinal structure function $F_L^{Q\bar{Q}}$ for $Q^2 \gg m^2$ were derived as the first result at $O(\alpha_s^3)$ [27].

The polarized deep inelastic parton densities are unfolded in QCD analyses of the structure function $g_1^{eN}(x, Q^2)$, presently at NLO, [28]. $\Delta u_v(x, Q^2)$ and $\Delta d_v(x, Q^2)$ are constrained best and an average statement can be made for the polarized sea under some assumptions as $SU(3)_F$ symmetry or fixed ratios among some of the sea quark distributions. Flavor tagged measurements were performed [29] to determine $\Delta q_i/q_i$ explicitely. These are first steps and higher luminosity measurements are required to reduce the errors further in the future. Constraints on the polarized gluon density are gotten through the QCD analysis and measuring open charm production. Yet the gluon density has a wide error band with mainly positive central values. First experimental results were obtained for the transversity structure function $h_1(x, Q^2)$ [30].

Important experimental tests concern the search for twist–3 contributions to polarized structure functions. For purely photonic interactions they are present in the structure function $g_2(x, Q^2)$ and, in the low Q^2 region, due to target mass effects, also in $g_1(x, Q^2)$ [31]. Similar to the Wandzura-Wilczek relation and other twist–2 relations in case of electro–weak interactions, the twist 3 contributions are related by integral relations, which can be tested in high luminosity experiments operating in the lower Q^2 region.

There is an ongoing programme to study deeply–virtual Compton scattering both at HERA and JLAB for a large set of observables. In course of these investigations one may hope to derive more information on the transverse sub–structure of nucleons and, potentially in the long term, information on parton angular momentum.

4. Future Avenues

Various important questions on the short–distance structure of nucleons are yet open and require further experimentation and more theoretical work. In the short run HERA will collect higher luminosity and measure $F_2(x, Q^2), F_2^{c\bar{c}}(x, Q^2)$ with much higher precision. Different experiments will yield more detailed results on $g_2(x, Q^2)$ and the transversity distribution $h_1(x, Q^2)$. One of the central issues is to measure $F_L(x, Q^2)$ with high accuracy in different ranges of Q^2 and it would be essential to perform this measurement at HERA due to its unique kinematic domain. Much of our understanding of the gluon distribution depends on this measurement.

RHIC and LHC will lead to improved constraints on the gluon and sea quark distributions both for polarized and unpolarized nucleons. JLAB will contribute with high precision measurements in the large x domain both for unpolarized and polarized nucleons, yet at low values of Q^2, which will increase with the advent of the higher energy option soon. These measurements supplement HERA's high precision measurements at small x. The possibility to investigate unpolarized and polarized deep inelastic scattering at the same experiments is crucial to minimize systematic errors. JLAB provides ideal facilities to experimentally explore twist–3 effects in g_2 at high precision and to extract higher twist effects for unpolarized structure functions in unified analyses including data from large virtuality DIS at CERN and HERA.

For the time after HERA different ep projects are discussed[a]. Two of the projects are ERHIC and ELIC. The kinematic regions for both proposals is situated between the domain explored at CERN and HERA before, cf. Figure 2. While ERHIC reaches somewhat lower values of x, ELIC will have the higher luminosity, increasing that of HERA by a factor of 1000 to 8000. As we saw before, various precision measurements are yet to be

---

[a]Future high-luminosity muon– and neutrino–factories will yield essential contributions to DIS, in particular concerning a detailed exploration of the sea-quark sector both for unpolarized and polarized nucleons. These projects are somewhat further ahead in time.

performed in this region. The programme will not concern inclusive quantities only but explore with sufficient luminosity also more rare channels, which are otherwise inaccessible but yield important theory tests. The high luminosity programme will answer many of the present open questions in the central kinematic region and yield challenging non-trivial tests of Quantum Chromodynamics both concerning perturbative and non–perturbative predictions. On the theory side, perturbative higher order calculations will continue at the level of $O(\alpha_s^3)$ corrections and heavy quark mass effects will be included. As much of the information during the next decade will come from proton–colliders, the respective processes have to be understood at higher precision. At the same time essential progress is expected in the systematic understanding of the moments of parton distributions on the lattice and for measuring the QCD scale Λ_{QCD}. In this way, high luminosity experimental results, high order perturbative calculations, and highly advanced non-perturbative techniques together will detail our understanding both of the strong force and the nature of nucleons at short and longer distances.

Particle physics always went along two avenues : i) the search for new particles in annihilation processes at ever increasing energies; ii) the search for new sub–structures of matter in resolving shorter and shorter distances, the atomic nucleus and finally the nucleons. At present it seems that quarks are point–like particles. Since this may be temporarily an impression, the search for their possible sub–structure has to be continued with suitable facilities in the future.

References

1. E.D. Bloom et al., Phys. Rev. Lett. **23** (1969) 930;
   M. Breidenbach et al., Phys. Rev. Lett. **23** (1969) 935.
2. M. Botje, M. Klein and C. Pascaud, `hep-ph/9609489`.
3. J.D. Bjorken, Phys. Rev. **179** (1969) 1574;
   K.G. Wilson, Phys. Rev. **179** (1969) 1699;
   R.A. Brandt and G. Preparata, Fortschr. Phys. **18** (1970) 249.
4. R.P. Feynman, Photon-Hadron Interactions, (Benjamin, Reading, MA, 1972).
5. D.J. Gross and S.B. Treiman, Phys. Rev. **D4** (1971) 1059.
6. H. Fritzsch and M. Gell-Mann, Proc. XVI Int. Conf on High Energy Physics, Fermilab, 1972, Vol. 2, pp.135.
7. D.J. Gross and F. Wilczek, Phys. Rev. Lett. **30** (1973) 1343;
   H.D. Politzer, Phys. Rev. Lett. **B30** (1973) 1346.
8. D.J. Gross and F. Wilczek, Phys. Rev. **D8** (1973) 3633; **D9** (1974) 980;
   H. Georgi and H.D. Politzer, Phys. Rev. **D9** (1974) 416.
9. W.L. van Neerven and E.B. Zijlstra, Nucl. Phys. **B383** (1992) 525 Nucl. Phys. **B417** (1994) 61. For an extensive list of references see Refs. [12a,d].

10. S.-O. Moch, J.A.M. Vermaseren, and A. Vogt, Nucl. Phys. **B688** (2004) 101; **B691** (2004) 129; Phys. Lett. **B606** (2005) 123; hep-ph/0504242.

11. J.A.M. Vermaseren, math-ph/0010025.

12. J. Blümlein and S. Kurth, Phys. Rev. **D60** (1999) 014018;
J. Vermaseren, Int. J. Mod. Phys. **A14** (1999) 2037;
J. Blümlein, Comp. Phys. Commun. **133** (2000) 76; **159** (2004) 19;
E. Remiddi and J. Vermaseren, Int. J. Mod. Phys. **A15** (2000) 725;
S.-O. Moch, P. Uwer, and S. Weinzierl, J. Math. Phys. **43** (2002) 3363.

13. E.A. Kuraev, L.N. Lipatov, and V.S. Fadin, Sov. Phys. JETP **45** (1977) 199;
M. Ciafaloni, Nucl. Phys. **B296** (1988) 49;
S. Catani and F. Hautmann, Nucl. Phys. **B 427** (1994) 475;
R. Kirschner and L.N. Lipatov, Nucl. Phys. **B213** (1983) 122.

14. V.S. Fadin and L.N. Lipatov, Phys. Lett. **B429** (1998) 127;
G. Camici and M. Ciafaloni, Phys. Lett. **B430** (1998) 349.

15. J. Blümlein and A. Vogt, Phys. Lett. **B370** (1996) 149; **B386** (1996) 350;
Phys. Rev. **D58** (1998) 014020;
J. Blümlein, V. Ravindran, W.L. van Neerven, and A. Vogt, hep-ph/9806368;
R. Ball and S. Forte, hep-ph/9805315.

16. J. Blümlein and H. Böttcher, Nucl. Phys. **A721** (2003) 333c.

17. F. Dittes, B. Geyer, J. Horejsi, D. Müller and D. Robaschik, Fortschr. Phys. **42** (1993) 2;
J. Blümlein, B. Geyer, and D. Robaschik, Nucl. Phys. **B560** (1999) 283;
A.V. Belitsky and A.V. Radyushkin hep-ph/0504030 and references therein.

18. X. Ji, Phys. Rev. Lett. **78** (1997) 610.

19. J. Blümlein and D. Robaschik, Phys. Lett. **B517** (2001) 222; Phys. Rev. **D65** (2002) 096002.

20. J. Blümlein, M. Klein, T. Naumann, and T. Riemann, PHE 88-01, Proc. of the HERA Workshop, Hamburg, 1987, Vol. 1, pp. 67;
J. Blümlein, G. Ingelman, M. Klein, and R. Rückl, Z. Phys. **C45** (1990) 501.

21. J. Blümlein, H. Böttcher, and A. Guffanti, hep-ph/0407089.

22. J. Blümlein and A. Guffanti, hep-ph/0411110.

23. S.A. Alekhin, Phys. Rev. **D68** (2003) 014002.

24. A.D. Martin et al., Eur. Phys. J. **C32** (2002) 73.

25. R.S. Towell et al., Phys. Rev. **D64** (2001) 052002.

26. E. Laenen, S. Riemersma, J. Smith and W.L. van Neerven, Nucl. Phys. **B392** (1993) 162;
S. Riemersma, J. Smith and W.L. van Neerven, Phys. Lett. **B347** (1995) 143.

27. J. Blümlein, A. De Freitas, S.-O. Moch, W.L. van Neerven, and S. Klein, in preparation.

28. Y. Goto et al., Phys. Rev. **D62** (2000) 034017; Int. J. Mod. Phys. **A18** (2003) 1203.
M. Glück et al., Phys. Rev. **D53** (2001) 094005;
J. Blümlein and H. Böttcher, Nucl. Phys. **B636** (2002) 225.

29. A. Airapetian et al., HERMES collaboration, hep-ex/0407032.

30. A. Airapetian et al., Phys. Rev. Lett. **94** (2005) 012002.

31. J. Blümlein and A. Tkabladze, Nucl. Phys. **B553** (1999) 427.

THE TRANSITION BETWEEN PERTURBATIVE AND NON-PERTURBATIVE QCD

A. FANTONI

Laboratori Nazionali di Frascati dell'INFN, Via E. Fermi 40,
00044 Frascati (RM), Italy

S. LIUTI

University of Virginia, Charlottesville, Virginia 22901, USA

We study both polarized and unpolarized proton structure functions in the kinematical region of large Bjorken x and four-momentum transfer of few GeV^2. In this
region the phenomenon of parton-hadron duality takes place between the smooth
continuation of the deep inelastic scattering curve and the average of the nucleon
resonances. We present results on a perturbative-QCD analysis using all recent
accurate data with the aim of extracting the infrared behavior of the nucleon
structure functions.

1. Introduction

Parton-hadron duality is generally defined as the similarity between
hadronic cross sections in the Deep Inelastic Scattering (DIS) region and
in the resonance region. It encompasses therefore a range of phenomena
where one expects to observe a transmogrification from partonic to hadronic
degrees of freedom, a question, the latter, at the very heart of Quantum
ChromoDynamics (QCD).

A number of experiments were conducted in the early days of QCD
where the onset of parton-hadron duality was observed as the equivalence
between the *continuation* of the smooth curve describing different observables from a wide variety of high energy and large momentum transfer reactions – structure functions, sum rules, $R(s)$ for $e^+e^- \rightarrow$ hadrons, heavy
meson decays... – and the same observables in the low energy region characterized by low final state invariant mass values and resonance structure.
A fully satisfactory theoretical description of this phenomenon, that became to be accepted as a "natural" feature of hadronic interactions, is still
nowadays very difficult to obtain. Recent progress both on the theoretical

163

and experimental side [1,2], has however both renovated and reinforced the hadronic physics community's interest in this subject as also demonstrated by the very lively discussions among participants in the present Workshop.

In our contribution to the Workshop, we present evidence that standard Perturbative QCD (PQCD) approaches in the large Bjorken x region might not be adequate to describe parton-hadron duality. In particular, by conducting an analysis of the most recent polarized and unpolarized inclusive electron scattering data, we unravel a discrepancy in the behavior of the extracted power corrections from the DIS and resonance regions, respectively.

2. Overview of data and QCD-based Interpretation

Parton-hadron duality being the idea that the outcome of any hard scattering process is determined by the initial scattering process among elementary constituents – the quarks and gluons – independently from the hadronic phase of the reaction, is a well rooted concept in our current view of all high energy phenomena. Bloom and Gilman (BG) duality [3] is the extension of this idea to a kinematical region characterized by lower center of mass energies of the hard scattering process. Recently, more accurate experimental data have been collected that allow us to explore in detail this phenomenon. Besides the already mentioned new data on both inclusive electron-proton scattering [1], and polarized inclusive electron-proton scattering [2], several additional data sets and hadronic reactions were measured in the resonance region: polarized inclusive electron-proton scattering at Jefferson Lab kinematics [4,5], $\tau \to \nu+$ hadrons [6], $\gamma p \to \pi^+ n$ [7,8], and, finally, inclusive electron-nucleus scattering [9] (most of the recent data were presented and discussed at this Workshop, see also [10] for a recent review of both theoretical and experimental results).

A particularly interesting result was found in studies of inclusive reactions with no hadrons in the initial state, such as $e^+e^- \to$ hadrons, and hadronic τ decays [6]. It was pointed out that, because of the truncation of the PQCD asymptotic series, terms including quark and gluon condensates play an increasing role as the center of mass energy of the process decreases. Oscillations in the physical observables were then found to appear if the condensates are calculated in an instanton background. Such oscillating structure, calculated in [6] for values of the center of mass energy above the resonance region, is damped at high energy, hence warranting the onset of parton-hadron duality.

In what follows we examine a related question, namely whether it is possible to extend the picture of duality explored in the higher Q^2 region [6], to the resonance region, or to the BG domain. A necessary condition is to determine whether the curve from the perturbative regime smoothly interpolates through the resonances, or whether, instead, violations of this correspondence occur. The latter would indicate that we are entering a semi-hard phase of QCD, where preconfinement effects might arise. Our analysis of "duality violations" requires a sufficiently large and accurate set of data. We have applied it therefore to both the unpolarized and polarized inclusive measurements of proton structure functions in the resonance region.

3. Monitoring the Transition between pQCD and npQCD

We outline two important procedures for the study of parton-hadron duality in structure functions: *i)* the *continuation* of DIS curve into the resonance region; *ii)* the *averaging* of the resonances. Although these concepts are equivalently found in a number of different reactions, and in different channels (see *e.g.* [10,6]), in this contribution we concentrate on the proton structure functions g_1 and F_2, for polarized, and unpolarized electron scattering, respectively.

3.1. *Continuation of DIS Curve*

It is important to define exactly what one means by "continuation" of the DIS curve, in order to be able to define whether parton-duality can be considered to be fulfilled. The accuracy of current data allows us, in fact, to address the question of *what extrapolation from the large Q^2, or asymptotic regime the cross sections in the resonance region should be compared to.* In principle any extrapolation from high to low Q^2 is expected to be fraught with theoretical uncertainties ranging from the propagation of the uncertainty on $\alpha_S(M_Z^2)$ into the resonance region to the appearance of different types of both perturbative and power corrections in the low Q^2 regime. All of these aspects need therefore to be evaluated carefully.

Our approach applies to the *large Bjorken x* behavior of inclusive data. We therefore consider:

(a) Non-Singlet (NS) Parton Distribution Functions (PDFs) evolved at Next to Leading Order (NLO);
(b) PQCD evolution using the scale $\approx Q^2(1 - z)$ which properly takes into account integration over the parton's transverse momentum [11,12];

(c) Target Mass Corrections (TMCs).

We perform an extensive study of inclusive data in the resonance region by extrapolating to this region all available parameterizations of PDFs, which can be considered pure DIS, down to the measured ranges for x, Q^2, and final state invariant mass, $W^2 = Q^2(1-x)/x + M^2$. Our general approach for both the unpolarized structure function $F_2(x, Q^2)$, and the polarized one, $g_1(x, Q^2)$ is described in detail in [14]. An important point illustrated also in [14] is that the uncertainty due to the use of different parameterizations can be taken into account by a band that is currently smaller than the experimental one in the region of interest. A potential theoretical error in the extrapolation of the ratios to low Q^2 could be generated by the propagation of the error in $\alpha_S(M_Z^2)$. However, because at large x perturbative evolution involves only Non-Singlet (NS) distributions, we expect it to affect minimally the extrapolation of the initial pQCD distribution, even to the low values of W^2 considered.

The problem of resumming the large logarithm terms arising at large x was first noticed in a pioneering paper [11]. There it was shown how this type of resummation can be taken care of by considering the correct definition of the upper limit of integration for the transverse momentum in the ladder diagrams defining the leading log approximation. This implies replacing Q^2 with $\approx \widetilde{W}^2 = Q^2(1-z)$, *i.e.* an invariant mass, in the evolution equations. Such a procedure was used to obtain our results both in Refs.[14,15], and in the current contribution.

Finally, TMCs are expected to be important at $W^2 \to M^2$. Although TMCs are kinematic in nature and their contribution to the Operator Product Expansion (OPE) was calculated early on [13], their effect on structure functions evolution cannot be evaluated straightforwardly at leading twist since a truncation of the twist expansion brings inevitably to a mismatch between the Bjorken x supports of the TM-corrected and the "asymptotic" results [16]. The extent of this mismatch is, however, small in the DIS region thus rendering approximated treatments applicable [13]. In the resonance region the discrepancies arising in the [13] approach can be large. We adopt, therefore, the prescription of Ref. [16], according to which the twist expansion for the standard moments Mellin of the structure function including TMCs is truncated consistently at the same order in $1/Q^2$ for both the kinematic terms and the dynamical higher twists. As noticed already in [15], this method applies for sufficiently small values of the expansion parameter $\approx 4M^2x^2/Q^2$. In addition, we have control on the uncertainty which is a term of $\mathcal{O}(1/Q^4)$.

3.2. Averaging Procedure

Resonant data can be averaged over, according to different procedures We considered the following complementary methods:

$$I(Q^2) = \int_{x_{\min}}^{x_{\max}} F_2^{\text{res}}(x, Q^2)\, dx \tag{1}$$

$$M_n(Q^2) = \int_0^1 dx\, \xi^{n-1}\, \frac{F_2^{\text{res}}(x, Q^2)}{x}\, p_n \tag{2}$$

$$F_2^{\text{ave}}(x, Q^2(x, W^2)) = F_2^{\text{Jlab}}(\xi, W^2) \tag{3}$$

where F_2^{res} is evaluated using the experimental data in the resonance region [a]. In Eq.(1), for each Q^2 value: $x_{\min} = Q^2/(Q^2 + W_{\max}^2 - M^2)$, and $x_{\max} = Q^2/(Q^2 + W_{\min}^2 - M^2)$. $W_{\min}$ and $W_{\max}$ delimit either the whole resonance region, i.e. $W_{\min} \approx 1.1$ GeV^2, and $W_{\max}^2 \approx 4$ GeV^2, or smaller intervals within it. In Eq.(2), ξ is the Nachtmann variable [17], and $M_n(Q^2)$ are Nachtmann moments [17]. The r.h.s. of Eq.(3), $F_2^{\text{Jlab}}(\xi, W^2)$, is a smooth fit to the resonant data [1], valid for $1 < W^2 < 4$ GeV^2; F_2^{ave} symbolizes the average taken at the $Q^2 \equiv (x, W^2)$ of the data.

4. Results

After describing our program to address quantitatively all sources of theoretical errors started in [15,14], in Fig.1 we present our main results on the extraction of the dynamical Higher Twist (HT) terms from the resonance region. A clear discrepancy marking perhaps a *breakdown of the twist expansion* at low values of W^2 is seen for the unpolarized structure function, F_2 (upper panel). A comparison with other results obtained in the DIS region [18] is also shown. For the polarized structure function, g_1, we added to our previous analysis data from [4] at $Q^2 = 0.65, 1, 1.2$ GeV^2. In addition, we used the experimental values of the ratio $R = \sigma_L/\sigma_T$ from recent Jefferson Lab measurements in the resonance region [21] which introduce an oscillation around the original result of about 2%, well within the error bars. A complete presentation and discussion of these results along with comparisons with other extractions [19,20] will be given in a forthcoming paper [22]. From

---

[a]Similar formulae hold for the polarized structure function, g_1.

the figure one can see that although the trend seen in [14] of a large violation of duality at $Q^2 \approx 1$ GeV^2 seems to be confirmed, more polarized data at large x are needed in order to draw definite conclusions.

This work is partially supported (S.L.) by the U.S. Department of Energy grant no. DE-FG02-01ER41200.

References

1. I. Niculescu *et al.*, Phys. Rev. Lett. **85**, 1186 (2000).
2. A. Airapetian *et al.* [HERMES Collaboration], Phys. Rev. Lett. **90**, 092002 (2003)
3. E.D. Bloom and F.J. Gilman, Phys. Rev. **D4**, 2901 (1971); Phys. Rev. Lett. **25**, 1140 (1970).
4. R. Fatemi *et al.*, Phys. Rev Lett. **91**, 222002 (2003).
5. Y. Prok, *these proceedings*.
6. M. A. Shifman, arXiv:hep-ph/0009131; I. I. Bigi and N. Uraltsev, Int. J. Mod. Phys. A **16**, 5201 (2001); *these proceedings*.
7. L.Y. Zhu *et al.*, Phys. Rev. Lett. **91** 0220043 (2003); Phys. Rev. **C71** 044603 (2005).
8. H. Gao, *these proceedings*.
9. J. Arrington *et al.*, nucl-ex/0307012.
10. W. Melnitchouk, R. Ent and C. Keppel, Phys. Rept. **406**, 127 (2005)
11. S. J. Brodsky and G. P. Lepage, SLAC-PUB-2447, *Presented at Summer Inst. on Particle Physics, SLAC, Stanford, Calif., Jul 9-20, 1979*
12. R. G. Roberts, Eur. Phys. J. C **10**, 697 (1999)
13. H. Georgi and H. D. Politzer, Phys. Rev. D **14**, 1829 (1976).
14. N. Bianchi, A. Fantoni and S. Liuti, Phys. Rev. D **69**, 014505 (2004)
15. S. Liuti, R. Ent, C. E. Keppel and I. Niculescu, Phys. Rev. Lett. **89**, 162001 (2002)
16. J. L. Miramontes and J. Sanchez Guillen, Z. Phys. C **41**, 247 (1988).
17. O. Nachtmann, Nucl. Phys. B **63**, 237 (1973).
18. A. D. Martin, R. G. Roberts, W. J. Stirling and R. S. Thorne, Eur. Phys. J. C **35**, 325 (2004)
19. E. Leader, A. V. Sidorov and D. B. Stamenov, Phys. Part. Nucl. Lett. **1**, 229 (2004), *and references therein*
20. D. B. Stamenov, *these proceedings*.
21. C. Keppel, *private communication*.
22. A. Fantoni and S. Liuti, *in preparation*.

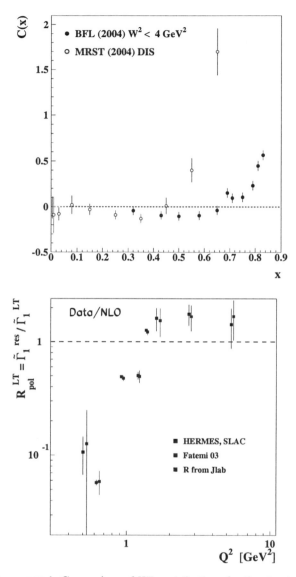

Figure 1. Upper panel: Comparison of HT contributions for the structure function F_2 in the DIS and resonance regions, respectively. The full circles are the values obtained in the resonance region [14]. For F_2 these are compared with extractions using DIS data, from [18]. Lower panel: ratio of the experimental data on g_1 [2,4] and the PQCD extrapolation to the resonance region [22]. Notice that the new results obtained using data from [4] agree with the trend of the Hermes data.

HIGHLY EXCITED HADRONS IN QCD AND BEYOND *

M. SHIFMAN

William I. Fine Theoretical Physics Institute, University of Minnesota,
Minneapolis, MN 55455, USA

I discuss two issues related to high "radial" excitations which attracted much attention recently: (i) chiral symmetry restoration in excited mesons and baryons, and (ii) universality of the ρ-meson coupling in QCD and AdS/QCD. New results are reported and a curious relation between an AdS/QCD formula and 1977 Migdal's proposal is noted.

1. Introduction

A renewed interest to highly excited mesons in QCD is explained by at least two reasons.

First, the gravity/gauge correspondence, originally established for conformal field theories on the gauge side, is being aggressively expanded to include closer relatives of QCD, with the intention to get a long-awaited theoretical control over QCD proper. The present-day "bottom-up" approach is as follows: one starts from QCD and attempts to guess its five-dimensional holographic dual. In this way, various holographic descriptions of QCD-like theories emerge and their consequences are being scrutinized with the purpose of finding the fittest model.

This approach goes under the name "AdS/QCD." Since the limit $N = \infty$ is inherent to AdS/QCD, the meson widths vanish, and one can unambiguously define masses and other static characteristics of excited states. Explorations of this type were reported in Refs. [1–17]. In refined holographic descriptions expected to emerge in the future one hopes to get asymptotically linear primary and daughter Regge trajectories, and obtain residues and other parameters and regularities pertinent to hadronic

---

*Based on invited talks delivered at the *First Workshop on Quark-Hadron Duality and the Transition to PQCD*, Laboratori Nazionali di Frascati, Italy, June 6–8, 2005, and the *Workshop on Highly Excited Hadrons*, ECT* in Trento, Italy, July 4–9, 2005.

physics, and demonstrate their compatibility with experiment — a goal which has not yet been achieved.

The recently suggested orientifold large-N expansion [18] which is complementary to the 't Hooft one [19], provides another framework which can be used for studying excited states.

The second reason is a clear-cut demonstration that the chiral symmetry of the QCD Lagrangian is empirically restored in excited mesons and baryons, due to Glozman and collaborators [20–23]. As is well-known from the early days of QCD, highly excited hadrons can be described quasiclassically (see e.g. [24]; recapitulated recently in Ref. [25] at the qualitative level). The quasiclassical description implies, in particular, asymptotically linear Regge trajectories. Needless to say, it also implies linear realization of all symmetries of the QCD Lagrangian at high energies, in particular, in high radial excitations. Indeed, in such states the valent quarks on average have high energies — high compared to Λ_{QCD} — and, thus, can be treated quasiclassically so that effects due to the quark condensate are inessential and can be neglected.

Nevertheless, the fact that excited states do exhibit the full linearly-realized chiral symmetry of QCD seemingly caught some theorists by surprise, probably, due to an over-concentration on the low-lying states for which the chiral symmetry is broken. (It would be more exact to say that the axial $SU(N_f)$ where N_f is the number of massless flavors is realized non-linearly, in the Goldstone mode.) According to [20–23], the symmetry is largely restored already for the first radial excitations; for instance, three excited pions and one excited scalar-isoscalar meson form an almost degenerate dimension-4 representation of $SU(2) \times SU(2) \sim O(4)$.[a]

The question is what can be said quantitatively on the rate of the symmetry restoration. In other words, what is the n-dependence (n is the excitation number) of the splittings δM_n of the radially excited states from one and the same representation of $SU(2) \times SU(2)$?

In the first part of my talk I suggest some natural "pedestrian" estimates

---

[a] I assume for simplicity that there are two massless flavors, ignoring the strange quark. The chiral U(1) also gets restored as a valid flavor symmetry: the U(1) chiral anomaly dies off in the limit of large number of colors. Even if the number of colors N is kept fixed, at large energies (i.e. high n), where perturbation theory becomes, in a sense, applicable, the axial U(1) gets restored since one can always redefine the quark U(1) current by adding a Chern-Simons current in such a way that the U(1) charge is conserved in perturbation theory. Restoration of the axial U(1) in radially excited mesons is clearly seen in experiment, cf. e.g. [22].

of the rate of the chiral symmetry restoration. In the second part I will focus on an aspect of AdS/QCD which was recently discussed in [13, 14]: implementation of the vector meson dominance (VMD) and universality of the ρHH coupling. AdS/QCD-based results will be confronted with QCD expectations.

2. Chiral symmetry restoration: generalities

If both vector and axial SU(2)'s are linearly realized, hadronic states must fall into degenerate multiplets of the full chiral symmetry. The degeneracy is lifted by an SU(2)×SU(2)→SU(2) breaking that dies off with energy (or the excitation number, which is the same). I will first briefly review an appropriate representation theory [22]. My next task will be an estimate of δM_n versus n at $n \gg 1$ where δM_n is the mass difference in particular chiral representations, for instance, the mass difference of the scalar–pseudoscalar mesons or the vector–axial-vector ones.

It is clear that to define highly excited states *per se* we need large N. In tricolor QCD, as we will shortly see, resonance widths rapidly grow with the excitation number; as a result, for large n the very separation of mesons from continuum becomes impossible, and the question of mass splittings and other similar questions cannot be addressed. *I will systematically exploit the large-N limit.* I will consider the massless quark sector consisting of two flavors. Extension to three flavors is rather straightforward.

The large-N limit sets an appropriate theoretical framework for consideration of excited mesons. As for excited baryons, there is no reason for their widths to be suppressed at $N \to \infty$. Therefore, theoretical analysis of highly excited baryonic states becomes problematic, as at $n \gg 1$ they should form a continuum. I will limit my discussion to mesonic states. (Empirically, isolation of excited baryon resonances from existing hadronic data seems possible, and data seem to indicate restoration of the full chiral symmetry for excited baryons [20–23]. Representations of unbroken SU(2)×SU(2) for baryons had been studied long ago, even before the advent of QCD [26].)

Analysis of the chiral symmetry is more conveniently performed in terms of the Weyl rather than Dirac spinors. Each Dirac spinor is a pair of two Weyl ones. We can take them both to be left-handed, χ and η, the first being triplet with respect to SU(3)$_{\text{color}}$, the second antitriplet.[b] To form

---

[b]If the gauge group is extended to SU(N), with respect to SU(N)$_{\text{color}}$ the field χ trans-

color singlets we convolute $\chi\eta$ or, alternatively, $\bar{\chi}\chi$, $\bar{\eta}\eta$. The Dirac spinor Ψ combines one left-handed and one right-handed spinor, $\Psi \sim \{\chi_\alpha, \bar{\eta}_{\dot{\beta}}\}$ where α and $\dot{\beta}$ are spinorial indices. Each chiral spinor carries a flavor SU(2) index. Since there are two linearly realized SU(2)'s, there are two flavor SU(2) indices, χ^k and $\eta^{\dot{a}}$ ($k, \dot{a}, = 1, 2$) which are independent. We can call them "left" and "right" isospin. Conventional isospin entangles "left" and "right" isospins.

The exact conserved quantum numbers of QCD, namely, conventional isospin, total angular momentum and parity, do not always completely specify the full structure of a quark-antiquark meson, as we will see shortly. Distinct patterns of "left" and "right" isospin additions can lead to distinct mesons having the same conventional isospin, total angular momentum and parity.

Let us now discuss the simplest representations. The scalar (pseudoscalar) mesons are of the type

$$\chi\eta \pm \bar{\chi}\bar{\eta}. \qquad (1)$$

Its Lorentz structure is $(1/2, 0) \times (1/2, 0) \to (0, 0)$ and $(0, 1/2) \times (0, 1/2) \to (0, 0)$. The isospin structure is $(1/2, 1/2)$. In terms of conventional isospin this is a triplet plus a singlet. In terms of $SU(2) \times SU(2) \sim O(4)$ we have two four-dimensional chiral representations: The first includes scalar isoscalar plus pseudoscalar isovector, the second pseudoscalar isoscalar plus scalar isovector.

In fact, the symmetry that gets restored is higher than just $SU(2) \times SU(2) \sim O(4)$. Indeed, at $N \to \infty$ the two-point functions of the currents $\bar{\Psi}\Psi$ and $\bar{\Psi}\tau^a\Psi$ are degenerate (here τ^a are the conventional isospin generators, $a = 1, 2, 3$), since the quark-gluon mixing that can occur in the isoscalar — but not isovector — channel is suppressed by $1/N$ and, thus, can be neglected. The above degeneracy is in one-to-one correspondence with the fact that the full flavor symmetry of the QCD Lagrangian is $U(1)_V \times U(1)_A \times SU(2) \times SU(2)$. The vector U(1), the baryon charge, plays no role in the meson sector. The linear realization of $U(1)_A \times SU(2) \times SU(2)$ implies that the two four-dimensional representations of $SU(2) \times SU(2)$ are combined in one irreducible eight-dimensional representation of $U(1)_A \times SU(2) \times SU(2)$.

If we pass to nonvanishing angular momenta, we observe that the vector

---

form as N, while η as $\overline{N}$.

and axial-vector mesons can be of two types,[c]

$$\bar{\chi}_{\dot{\alpha}}\chi_\alpha \pm \bar{\eta}_{\dot{\alpha}}\eta_\alpha \,, \tag{2}$$

$$\chi_{\{\alpha}\eta_{\beta\}} \pm \text{h.c.} \,, \tag{3}$$

where the braces denote symmetrization. The first one is Lorentz $(1/2, 1/2)$, the second $(1, 0) + (0, 1)$. In terms of the "left" and "right" isospins, it is the other way around. The state $\bar{\chi}_{\dot{\alpha}}\chi_\alpha$ has isospin $(\overline{1/2}, 0) \times (1/2, 0) \to (1, 0) + (0, 0)$. It is a triplet plus a singlet with respect to the conventional isospin. Taking into account the second term in (2), we have a vector isovector plus an axial-vector isovector plus two isosinglets. At large N they all, taken together, form an eight-dimensional representation. The state (3) forms SU(2)×SU(2) quadruplets $(1/2, 0) \times (0, 1/2)$: a vector isoscalar plus an axial-vector isovector, and vice versa, a vector isovector plus an axial-vector isoscalar. Again, both quadruplets are combined into an eight-dimensional representation at $N \to \infty$.

A physical ground-state ρ meson which is roughly an equal mixture of (2) and (3) is "polygamous" and has two distinct chiral partners [22]: an axial-vector isovector and an axial-vector isoscalar.

3. The rate of the chiral symmetry restoration

For simplicity I will discuss the mass splittings of scalar versus pseudoscalar excited mesons, produced from the vacuum by the operators $\bar{\Psi}\Psi$ and $i\bar{\Psi}\gamma_5\Psi$, respectively. As was mentioned, at $N \to \infty$ their isotopic structure is inessential. The mass splittings in other chiral multiplets must have the same n dependence.

The chiral symmetry is broken by the quark mass term, which I will put to zero. Then the local order parameter representing the chiral symmetry breaking is $\langle\bar{\Psi}\Psi\rangle$. Unfortunately, it is rather hard to express the mass splittings in the individual multiplets in terms of this local parameter. Generally speaking, the fact that its mass dimension is 3 tells us that various chiral symmetry violating effects will die off as inverse powers of M_n (i.e. as positive powers of $n^{-1/2}$). Today's level of command of QCD does not allow us to unambiguously predict the laws of fall off of the chiral symmetry violating effects in terms of $\langle\bar{\Psi}\Psi\rangle$. In some instances a minimal

---

[c]This fact was, of course, known to QCD practitioners from the very beginning, and was repeatedly exploited in QCD and in models in various contexts, e.g. [27].

rate can be estimated, however. In some instances there are reasons to believe that the actual rate may be close to the minimal one.

3.1. On quasiclassical arguments

As a warm-up exercise I will derive textbook results: equidistant spacing of radially excited mesons and n independence of Γ_n/M_n where Γ_n is the total width of the n-th excitation. In the following simple estimates I will try to be as straightforward as possible, omitting inessential numerical constants and assuming that the only mass dimension is provided by $\Lambda_{\text{QCD}} \equiv \Lambda$. In this "reference frame" the string tension is Λ^2, while the ρ-meson mass is Λ.

When a highly excited meson state (say, ρ_n) is created by a local source (vector current), it can be considered, quasiclassically, as a pair of free ultrarelativistic quarks; each of them with $E = p = M_n/2$. These quarks are produced at the origin, and then fly back-to-back, eventually creating a flux tube of the chromoelectric field. Since the tension of the flux tube is $\sigma \sim \Lambda^2$, the length of the tube is

$$L \sim \frac{M_n}{\Lambda^2}. \tag{4}$$

Using the quasiclassical quantization condition

$$\int p\,dx \sim p\,L \sim \frac{M_n^2}{\Lambda^2} \sim n \tag{5}$$

we immediately arrive at

$$M_n^2 \sim \Lambda^2\, n. \tag{6}$$

(Let me parenthetically note that asymptotically linear n dependence of M_n^2 can be analytically obtained in the two-dimensional 't Hooft model [28,29] where linear confinement is built in.)

Let us now discuss total decay widths of high radial excitations. The decay probability (per unit time) is determined, to order $1/N$, by the probability of producing an extra quark-antiquark pair. Since the pair creation can happen anywhere inside the flux tube, the probability must be proportional to L. As a result [30],

$$\Gamma_n \sim \frac{1}{N}\,L\Lambda^2 = \frac{B}{N}\,M_n, \tag{7}$$

where B is a dimensionless coefficient independent of N and n, see Eq. (4).

Thus, the width of the n-th excited state is proportional to its mass which, in turn, is proportional to $\sqrt{n}$. The square root formula for Γ_n was numerically confirmed [31] in the 't Hooft model. It is curious that both, in actual QCD and in two dimensions, $B \sim 0.5$.

Equation (7) demonstrates that the limits $N \to \infty$ and $n \to \infty$ are not commutative. We must first send N to infinity, and only then can we consider high radial excitations.

Asymptotic linearity of the Regge trajectories (Eq. (6) at $n \gg 1$) must emerge in any sensible string-theory-based description of QCD. As discussed in detail by Schreiber [32], this is indeed the case in the picture of mesons as open strings ending on a probe D brane in an appropriate background. In the same work, using an open string analog of the well-known Witten's argument, Shreiber shows [32] that treating radial excitation of low-spin mesons as fluctuations of the probe D branes one obtains, generally speaking, a wrong behavior, $M_n^2 \sim \Lambda^2 n^2$. (This remark which pre-emps the beginning of Sect. 4 will be explained there in more detail.)

3.2. Analyzing chiral pairs: the slowest fall off of δM_n

For definiteness I will focus on scalar–pseudoscalar mesons. Semi-quantitative results to be derived below are straightforward and general. The only "serious" formula I will use is that for the Euler function,

$$\psi(z) = -\gamma - \frac{1}{z} + \sum_{n=1}^{\infty} \left(\frac{1}{n} - \frac{1}{z+n} \right). \tag{8}$$

At large positive z

$$\psi(z) \to \ln z - \frac{1}{2z} + O(z^{-2}). \tag{9}$$

The two-point correlators we will deal with are defined as

$$\Pi(q) = i \int d^4 x e^{i q x} \langle T\{J(x), J(0)\} \rangle, \tag{10}$$

where

$$J = \bar{\Psi}\Psi \text{ and } J_5 = i \bar{\Psi}\gamma^5 \Psi \tag{11}$$

for scalars and pseudoscalars, respectively (to be denoted as as Π and Π_5). I will consider flavor-nonsinglet channels. Then

$$\Pi(Q) - \Pi_5(Q) = \sum_n \left(\frac{f_n}{Q^2 + M_n^2} - \frac{\tilde{f}_n}{Q^2 + \tilde{M}_n^2} \right), \tag{12}$$

where the untilded quantities refer to the scalar channel while those with tildes to the pseudoscalar one.

The operator product expansion (OPE) for $\Pi(Q) - \Pi_5(Q)$ at large Euclidean Q^2 was built long ago [33]. In conjunction with the large-N limit which justifies factorization of the four-quark operators it implies

$$\Pi(Q) - \Pi_5(Q) \sim \frac{1}{N} \frac{\langle \bar{\Psi}\Psi \rangle^2}{Q^4}, \qquad Q^2 \to \infty, \tag{13}$$

(modulo possible logarithms). Now, the residues f_n are positive numbers of dimension Λ^4. They can be normalized from the relation

$$\Pi(Q) \sim NQ^2 \ln Q^2, \qquad Q^2 \to \infty. \tag{14}$$

In fact, it is easy to show that the equidistant spectrum (6) combined with Eq. (8) lead to

$$f_n \sim N\Lambda^2 M_n^2. \tag{15}$$

Now we are in position to estimate the splittings. As was explained above, one expects that asymptotically, at large n,

$$|\delta f_n| \ll f_n, \qquad |\delta M_n^2| \ll M_n^2, \tag{16}$$

where

$$\delta f_n = f_n - \tilde{f}_n, \qquad \delta M_n^2 = M_n^2 - \tilde{M}_n^2. \tag{17}$$

The scalar-pseudoscalar difference in Eq. (12) depends on δf_n and δM_n^2. We first set $\delta f_n = 0$, i.e. assume perfectly degenerate residues. (Shortly this degeneracy will be lifted, of course.) Then, taking account of (15), we get the following sum-over-resonances representation:

$$\Pi(Q) - \Pi_5(Q) = N\Lambda^2 Q^2 \sum_n \frac{\delta M_n^2}{(Q^2 + M_n^2)^2}$$

$$\to N\Lambda^2 Q^2 \frac{\partial}{\partial Q^2} \sum_n \frac{\delta M_n^2}{Q^2 + M_n^2}. \tag{18}$$

To evaluate the convergence of δM_n^2 to zero, the last expression must be matched with the asymptotic formula (13) at $Q^2 \to \infty$. Needless to say, matching an infinite sum to a single OPE term one cannot expect to get a unique solution for δM_n^2. In fact, what we are after, is the *slowest* pattern of the chiral symmetry restoration still compatible with the OPE. [d]

---

[d] As we will see in Sect. 4, the advent of holographic ideas in QCD definitely revived interest in combining OPE with Regge-trajectory-based constructions. For instance,

Even if δM_n^2 falls very fast at large n, but the sign of δM_n^2 is n-independent, there is no matching. Indeed, Eq. (18) implies then $1/Q^2$ rather than $1/Q^4$ behavior. The only way to enforce the $1/Q^4$ asymptotics is to assume that δM_n^2 is a sign alternating function of n, say,

$$\delta M_{2k}^2 > 0, \qquad \delta M_{2k+1}^2 < 0, \qquad |\delta M_{2k}^2| = |\delta M_{2k+1}^2|. \qquad (19)$$

Let us show that the slowest possible decrease of $|\delta M_n^2|$ versus n is

$$M_n^2 \delta M_n^2 = \text{sign alternating const}. \qquad (20)$$

Equation (13) implies that $\delta \Pi(Q^2)$ is analytic in the complex plane, with a Q^{-4} singularity at the origin. This implies, in turn, that the sum

$$\sum_n \delta M_n^2 \qquad (21)$$

is convergent and vanishes. The slowest-rate solution is $\delta M_n^2 \sim (-1)^n n^{-1}$, which is the same as (20). Under the condition (20) the right-hand side of (18) becomes

$$N \Lambda^6 Q^2 \frac{\partial}{\partial Q^2} \sum_k \left\{ \frac{1}{M_{2k}^2 \left(Q^2 + M_{2k}^2\right)} - \frac{1}{M_{2k+1}^2 \left(Q^2 + M_{2k+1}^2\right)} \right\}$$

$$\to N \Lambda^4 Q^2 \frac{\partial}{\partial Q^2} \frac{1}{Q^2} \ln \frac{Q^2 + \Lambda^2}{Q^2} \to N \Lambda^6 \frac{1}{Q^4}, \qquad (22)$$

quod erat demonstrandum. Here Λ^6 must be identified with $N^{-2} \langle \bar{\Psi}\Psi \rangle^2$.

Thus, δM_n is sign alternating and falls off as

$$|\delta M_n| \sim \frac{\Lambda}{n^{3/2}} \qquad (23)$$

or *faster.* Whether or not it actually falls off faster than $n^{-3/2}$ depends on the behavior of higher order terms in OPE for $\Pi(Q) - \Pi_5(Q)$. Next-to-nothing can be said on this at the moment. Even with the slowest possible regime (23) the rate of the chiral symmetry restoration is pretty fast, see Fig. 1.

---

attempts to reconcile OPE in the chirality-breaking two-point function $\langle V + A, V - A \rangle$ with strictly linear Regge trajectories were reported in [34]. Of course, no reconciliation can be achieved under the assumption of exact linearity. Similar arguments were later used in [35] to address the issue of asymptotic deviations of the Regge trajectories from linearity.

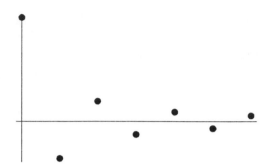

Figure 1. δM_n versus n, in arbitrary units.

3.3. The impact of $\delta f_n \neq 0$

Now, let us relax the (unrealistic) assumption $\delta f_n = 0$. Again, our task is to find the slowest fall off compatible with both the OPE and resonance representations. If both $\delta f_n \neq 0$ and $\delta M_n \neq 0$, the difference $\Pi(Q) - \Pi_5(Q)$ acquires an additional term

$$\Pi(Q) - \Pi_5(Q) = \sum_n \frac{\delta f_n}{Q^2 + M_n^2} . \tag{24}$$

It is not difficult to see that, barring an unlikely and subtle conspiracy between δf_n and δM_n, the previous estimate for δM_n stays intact, and, in addition, one gets an estimate for δf_n. Copying the consideration above one finds that δf_n must be sign-alternating, and the fall-off regime of $|\delta f_n|$ must be $N\Lambda^4/n$ or faster, so that

$$|\delta f_n|/f_n \sim 1/n^2 . \tag{25}$$

Here I used the fact that f_n grows linearly with n, see Eq. (15).

3.4. Excited ρ mesons of two kinds

In Ref. [22] it was noted that, if the chiral symmetry is restored at high n, two distinct varieties of excited ρ mesons must exist. Let us define ρ meson as a $J^{PC} = 1^{--}$ quark-antiquark state with isospin 1. Then, such mesons are produced from the vacuum by the following two currents [e] which belong

---

[e]The current (27) also produces 1^+ mesons; this is irrelevant for what follows.

to two distinct chiral multiplets (see Eqs. (2) and (3)):

$$\bar{\Psi}\vec{\tau}\gamma_\mu\Psi \,, \tag{26}$$

$$\bar{\Psi}\vec{\tau}\sigma_{\alpha\beta}\Psi \,. \tag{27}$$

Were the chiral symmetry exact, the above two currents would not mix. The ground state ρ meson would be coupled to the first current and would have vanishing residue in the second. There will be two distinct types of the $J^{PC} = 1^{--}$ mesons. In actuality, they do mix, however; the ρ mesons show up in both currents, see Fig. 2. The $J^{PC} = 1^{--}$ excited states of the first kind couple predominantly to the first current while those of the second kind to the second current. I want to evaluate the coupling of ρ_n of a given kind to the "wrong" current (as a function of n).

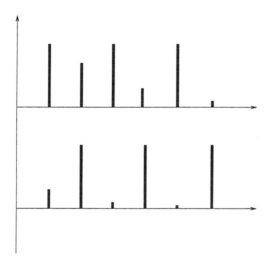

Figure 2. Spectral densities in the channels corresponding to the currents (26) and (27). Contributions due to ρ_n of the "wrong kind" die off as $1/n^3$ at large n.

To this end we will consider a "mixed" correlation function

$$\left\langle \bar{\Psi}\vec{\tau}\gamma_\mu\Psi \,,\ \bar{\Psi}\vec{\tau}\sigma_{\alpha\beta}\Psi \right\rangle_q \sim (g_{\alpha\mu}\, q_\beta - g_{\beta\mu}\, q_\alpha)\, \langle \bar{\Psi}\Psi \rangle \frac{1}{q^2} \tag{28}$$

modulo logs of Q^2. Saturating this expansion by resonances and defining

$$\langle \mathrm{vac}|\bar{\Psi}\vec{\tau}\sigma_{\alpha\beta}\Psi|\rho_n\rangle = (g_{\alpha\mu}\, p_\beta - g_{\beta\mu}\, p_\alpha)\, f_n\, \varepsilon_\mu^a \tag{29}$$

we get, say, for the resonances "strongly" coupled to $\Psi\vec{\tau}\gamma_\mu\Psi$

$$\sum_n \frac{M_n^2}{g_n} f_n \sim \langle\bar{\Psi}\Psi\rangle\,,\tag{30}$$

implying, in turn, that

$$f_n \sim \frac{\Lambda}{n^{3/2}}\tag{31}$$

or faster. Thus, barring conspiracies, the contribution, of an excited ρ of the "wrong" chiral structure in the *diagonal* two-point function of two vector currents $\bar{\Psi}\vec{\tau}\gamma_\mu\Psi$ scales as $f_n^2 \sim 1/n^3$, to be compared with that of the "right" chiral structure, $M_n^2/g_n^2 \sim n^0$. The relative suppression is $1/n^3$ or faster.

One can arrive at the very same statement using a slightly different argument. The wrong-chirality ρ_n's in the *diagonal* two-point function of two vector currents $\bar{\Psi}\vec{\tau}\gamma_\mu\Psi$ must be associated with the operator $\langle\bar{\Psi}\Psi\rangle^2$. Therefore, the sum $\sum_n f_n^2 M_n^4$ must converge, again leading to the estimate $f_n^2 \sim 1/n^3$ or faster.

4. AdS/QCD versus QCD

As I have already mentioned, AdS/CFT correspondence [36] inspired a general search for five-dimensional holographic duals of QCD. This trend flourished in the last few years, e.g. [1–17]. Generic holographic models of QCD describe gauge theories that typically (but not always!) have color confinement and chiral symmetry breaking as built-in features, and are dual to a string theory on a weakly-curved background. One should remember that they describe strong coupling theory — i.e. asymptotic freedom is not properly incorporated. In the majority of the holographic duals discussed so far $M_n \sim n$ rather than $\sqrt{n}$ required by linearity of the Regge trajectories. In some more contrived holographic descriptions (e.g. [16]) asymptotically linear Regge trajectories do emerge, however.

In discussing flavor physics in AdS/QCD one should keep in mind that at present there are two methods of introducing dynamical quarks in the fundamental representation into the gravity/string duals of QCD: (i) the so-called flavor probes (for $N_f \ll N$); (ii) fully back-reacted backgrounds that incorporate flavor — in this case, obviously, there are no restrictions on N_f. The latter direction is not yet sufficiently developed. Burrington et al. [9] suggested a background based on D3/D7 system, which seems singular, however. This construction has some undesirable features, serious drawbacks, which cannot be discussed here. Klebanov and Maldacena incorporated flavor [10] in a conformal theory claimed to be dual to the infrared fixed point of $N = 1$ SQCD, a theory which is in no way close to QCD. This seems to exhaust the list of developments in this direction.

The probe approach pioneered by Karch and Katz [1] is way more advanced. In the original work [1] Karch and Katz added D7 branes to the $AdS_5 \times S^5$ background and obtained a model which had no confinement. A remedy was found shortly. Flavor in a confining background was introduced in Ref. [4] in the context of the Klebanov-Strassler model. A simpler and more illuminating paper [5] followed almost immediately. It was based on adding D6 flavor branes to the model of Witten of near-extremal D4 branes. Although Ref. [5] was very inspiring, the authors themselves were aware of the fact this model did not incorporate chiral symmetry in the ultraviolet and, hence, dynamical chiral symmetry breaking of QCD was not addressed.

Dedicated designs allowing one to include chiral symmetry breaking and related low-energy phenomena in AdS/QCD were worked out in [6,8,11,15]. In fact, it will be very interesting to check whether fluctuations of the probe branes that correspond to meson excitations will explain chiral symmetry restoration in high excitations, and if yes, whether the rate of restoration in highly excited mesons will be properly reproduced.

This last remark presents a nice bridge between the contents of Sect. 3 and the remainder of this talk devoted to implementations of universality. First AdS/QCD-based analyses of the issue were reported in Refs. [13, 14]. After a brief review I confront them with QCD expectations which I derived for this conference. One should note that, notwithstanding their stimulating character, the models [13, 14] are not based on backgrounds and probes proven to be duals of QCD. In fact, in these models the Regge trajectories are not asymptotically linear. Moreover, they share a general feature inherent to all probe-based constructions. Ignoring back reaction means that the flavor-carrying quarks in these models are, in essence, non-dynamical. A conceptual parallel that immediately comes to one's mind is quenching in lattice QCD. This does not seem to be a serious drawback, though, given that $N = \infty$.

One could say that unless the above stumbling blocks are eliminated there is no point in analyzing consequences of present-day AdS/QCD and confronting them with QCD proper. Such a standpoint, although legitimate, is not constructive. The more aspects we study the more chances we have for eliminating drawbacks and finding a "perfectly good" holographic dual.

4.1. *Implementation of universality in AdS/QCD and QCD*

To begin with, I would like to dwell on the work of Hong et al. [13] devoted to the issue of VMD and universality of the ρ-meson coupling.

The notion of VMD is known from the 1960's [37]. Let us consider, for definiteness, the vector isovector current

$$J^\mu = \frac{1}{2} \left(\bar{u}\gamma^\mu u - \bar{d}\gamma^\mu d \right) . \tag{32}$$

If it is completely saturated by the ρ meson,

$$J^\mu \equiv \left(\frac{M_\rho^2}{g_\rho} \right) \rho^\mu , \tag{33}$$

with no higher excitations, then the $\rho H H$ coupling is obviously universal and depends only on the isospin of the hadron H. Indeed, consider the formfactor of H for the J^μ-induced transition. At zero momentum it equals to the isospin of H. On the other hand, saturating the formfactor by the ρ meson, we get

$$(g_{\rho H H}) \, g_\rho^{-1} = H \ \text{isospin} . \tag{34}$$

This is the famous VMD formula. Say, for the pion, $g_{\rho\pi\pi} = g_\rho$. In what follows I will keep in mind $H = \pi$ as a typical example.[f]

Equation (34) is approximate since the absolute saturation (33) is certainly unrealistic. Higher radial excitations are coupled to the isovector current too,

$$\langle \text{vac} | J^\mu | \rho_n \rangle = \left(\frac{M_n^2}{g_n} \right) \epsilon_n^\mu \neq 0 , \qquad n = 1, 2, 3, ... \tag{35}$$

The exact formula replacing VMD is

$$(g_{\rho H H}) \, g_\rho^{-1} + \sum_{n=1}^{\infty} \frac{g_{\rho_n H H}}{g_n} = H \ \text{isospin} . \tag{36}$$

If the sum over excitations on the left-hand side is numerically small, for whatever reason, one still recovers the universality relation (34) which will be valid approximately rather than exactly.

---

[f]In this case the (dimensionless) coupling constant $g_{\rho\pi\pi}$ is defined as

$$\langle \rho | \pi^+ \pi^- \rangle = g_{\rho\pi\pi} \, \epsilon_\mu \left(p_+^\mu - p_-^\mu \right) .$$

There are two distinct regimes ensuring suppression of the sum: (a) each term $n = 1, 2, 3, \ldots$ is individually small; (b) each term is of the same order as $(g_{\rho HH}) \, g_\rho^{-1}$, but successive terms are sign-alternating and compensate each other. The first option is an approximate VMD and, hence, leads to a natural universality, while the second one is in fact an "accidental" or " fortuitous" universality.

It is the latter regime which takes place in the holographic model considered in Ref. [13], see Fig. 3 illustrating numerical results reported in this paper.[g] For the ground state the authors get [13] $(g_{\rho HH}) \approx 1.49 \, g_\rho$. The factor 1.49 is to be compared with unity in VMD.

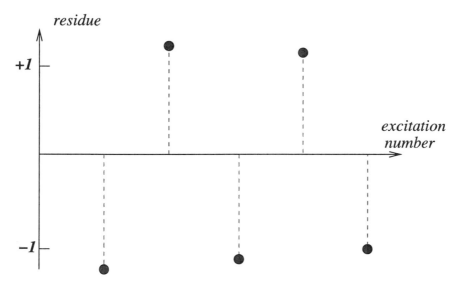

Figure 3.   The residue $g_{\rho_n HH} \, (g_n)^{-1}$ versus n for $n = 1, 2, \ldots$ Borrowed from [13].

I would like to show that QCD proper gives the former regime rather than the latter — suppression of individual contributions of high excitations. The divergence between QCD and AdS/QCD must cause no surprise

---

[g]Whether or not poor convergence of $(g_{\rho_n HH}) \, g_{\rho_n}^{-1}$ is a general feature of AdS/QCD, or, perhaps, of a certain class of holographic models, remains to be seen. According to M. Stephanov's private communication, the AdS/QCD model of Ref. [14], which builds on previous results [6, 8], yields $n^{-5/2}$, the rate of fall-off that is even steeper than that expected in QCD, see Sect. 4.2. An intermediate regime, with the fall-off rate $\sim n^{-1/2}$, is reported in [38]. It would be instructive to discuss in detail particular reasons explaining the above distinctions.

186

since AdS/QCD does not accurately describe short-distance QCD dynamics which governs high excitations.

Now, let us discuss QCD-based expectations in more detail.

4.2. *Sign alternating residues*

As well-known (see e.g. [39] for extensive reviews), at large (Euclidean) momentum transfer the pion formfactor is determined by the graph of Fig. 4 and scales as

$$F_\pi(Q^2) \sim \frac{1}{Q^2 \ln Q^2} \, . \tag{37}$$

It is important that the fall off is *faster* than $1/Q^2$. Comparing Eq. (37)

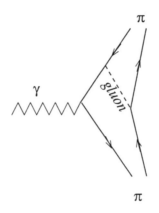

Figure 4.   Asymptotic behavior of the pion formfactor in QCD.

with the sum-over-resonances representation,

$$F_\pi(Q^2) = \sum_{n=0}^{\infty} \frac{g_{\rho_n \pi\pi}}{g_n} \frac{M_{\rho_n}^2}{Q^2 + M_{\rho_n}^2} \, , \tag{38}$$

we immediately conclude that successive terms must be sign-alternating — otherwise the asymptotic fall off would be $1/Q^2$. Thus, QCD supports the sign-alternation feature of AdS/QCD.

4.3. *Large n suppression of the residues*

Now, our task is to estimate the large n behavior of $g_{\rho_n \pi\pi}/g_n$ using, as previously, the quasiclassical approximation. A high radial excitation of the ρ meson can be viewed as a an ultrarelativistic quark-antiquark system, each quark having energy $m_n/2$ Conversion to the pion pair proceeds via the diagram of Fig. 5. A straightforward examination of this graph allows

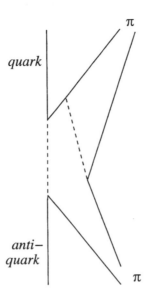

Figure 5.   Determination of $g_{\rho_n \pi \pi}$ at large n from two-gluon exchange graph.

us to conclude that

$$g_{\rho_n \pi \pi} \sim \frac{f_\pi^2}{g_n} \frac{1}{M_n^2} \frac{1}{[\ln(M_n^2/\Lambda^2)]^2} , \qquad (39)$$

where both factors, M_n^{-2} and the square of the logarithm, are the consequences of the two-gluon exchange. The logarithm is due to α_s (see below).

Deviations from VMD in Eq. (36) are determined by the residues

$$\frac{g_{\rho_n \pi \pi}}{g_n} \sim \frac{f_\pi^2}{g_n^2} \frac{1}{M_n^2} \frac{1}{[\ln(M_n^2/\Lambda^2)]^2} . \qquad (40)$$

Taking into account the fact that M_n^2/g_n^2 is asymptotically n-independent, we arrive at the following suppression factor:

$$\left| \frac{g_{\rho_n \pi \pi}}{g_n} \right| \sim \frac{1}{n^2} \frac{1}{(\ln n)^2} . \qquad (41)$$

the fall off of the residues of high excitations is quite steep. Formally, Eq. (41) is valid at large n. It is reasonable to ask whether a suppression persists for the first or second excitation. The answer seems to be positive. Even if $n \sim 1$, there is a numeric suppression coming from α_s^2 of the type

$$(\lambda)^2 = \left(\frac{N\alpha_s}{2\pi}\right)^2 \sim (4 \ln M_n/\Lambda)^{-2} , \qquad (42)$$

where λ is the 't Hooft coupling, small in QCD and large in AdS/QCD. According to the above estimate, the first excitation contributes in Eq. (36) at the 10% level. This is in accord with the experience I gained from multiple analyses of the QCD sum rules [33] in which I had been involved in the past.

5. In conclusion...

I would like to conclude this talk on a curious note showing that, perhaps, indeed, new is a well-forgotten old. Descending down from conceptual summits to down-to-earth technicalities let us ask ourselves what we learn from AdS/QCD with regards, say, to the ρ-meson channel at operational level. To this end let us have a closer look at results reported in [14]. Operationally, the bare-quark-loop logarithm is represented in this work as an infinite sum over excited ρ mesons whose masses and residues are adjusted in such a way that the above infinite sum reproduces pure logarithm of Q^2 up to corrections exponentially small at large Q^2 (there are no power corrections).

A similar question was raised long ago, decades before AdS/QCD, by A. Migdal [40], who asked himself what is the best possible accuracy to which $\log Q^2$ can be approximated by an infinite sum of infinitely narrow discrete resonances, and what are the corresponding values of the resonance masses and residues. He answered this question as follows: "the accuracy is exponential at large Q^2 and the resonances must be situated at the zeros of a Bessel function." This is exactly the position of the excited ρ mesons found in Ref. [14]!

6. Conclusions

◈ Chiral symmetry restoration in high radial excitations occurs at the rate

$$|\delta M_n| \sim \Lambda n^{-3/2}$$

or faster (this is related to the quark condensate $\langle \bar{\Psi}\Psi \rangle^2$);

◈◈ Relatively simple versions of AdS/QCD (the majority of the holographic duals analyzed so far!), along with the wrong n dependence of M_n (the well-known fact) also lead to a wrong pattern for the n dependence of the residue $g_{\rho_n \pi\pi}/g_n$. Thus, although the $g_{\rho HH}$ universality is implemented in AdS/QCD, it may be implemented in a wrong way!

Acknowledgments

I am grateful to T. Cohen, L. Glozman, T. Son, M. Stephanov, and A. Vainshtein for illuminating discussions. Very useful communications with A. Armoni, S. Beane, M. Chizhov, N. Evans, J. Hirn, D. Jido, I. Kirsch, J. Sonnenschein, and M. Strassler are gratefully acknowledged. This work was supported in part by DOE grant DE-FG02-94ER408.

References

1. A. Karch and E. Katz, JHEP **0206**, 043 (2002) [hep-th/0205236]; see also A. Karch, E. Katz and N. Weiner, Phys. Rev. Lett. **90**, 091601 (2003) [hep-th/0211107].
2. H. Boschi-Filho and N. R. F. Braga, Eur. Phys. J. C **32**, 529 (2004) [hep-th/0209080].
3. M. Kruczenski, D. Mateos, R. C. Myers and D. J. Winters, JHEP **0307**, 049 (2003) [hep-th/0304032].
4. T. Sakai and J. Sonnenschein, JHEP **0309**, 047 (2003) [hep-th/0305049].
5. M. Kruczenski, D. Mateos, R. C. Myers and D. J. Winters, JHEP **0405**, 041 (2004) [hep-th/0311270].
6. J. Babington, J. Erdmenger, N. J. Evans, Z. Guralnik and I. Kirsch, Phys. Rev. D **69**, 066007 (2004) [hep-th/0306018].
7. S. Hong, S. Yoon and M. J. Strassler, JHEP **0404**, 046 (2004) [hep-th/0312071].
8. N. J. Evans and J. P. Shock, Phys. Rev. D **70**, 046002 (2004) [hep-th/0403279].
9. B. A. Burrington, J. T. Liu, L. A. Pando Zayas and D. Vaman, JHEP **0502**, 022 (2005) [hep-th/0406207].
10. I. R. Klebanov and J. M. Maldacena, Int. J. Mod. Phys. A **19**, 5003 (2004) [hep-th/0409133].
11. T. Sakai and S. Sugimoto, Prog. Theor. Phys. **113**, 843 (2005) [hep-th/0412141]; *More on a holographic dual of QCD*, hep-th/0507073.
12. G. F. de Teramond and S. J. Brodsky, Phys. Rev. Lett. **94**, 201601 (2005) [hep-th/0501022].
13. S. Hong, S. Yoon and M. J. Strassler, *On the couplings of the ρ meson in AdS/QCD*, hep-ph/0501197.

14. J. Erlich, E. Katz, D. T. Son and M. A. Stephanov, *QCD and a holographic model of hadrons*, hep-ph/0501128.

15. L. Da Rold and A. Pomarol, *Chiral symmetry breaking from five-dimensional spaces*, hep-ph/0501218; H. Boschi-Filho, N. R. F. Braga and H. L. Carrion, *Glueball Regge trajectories from gauge/string duality and the Pomeron*, hep-th/0507063.

16. M. Kruczenski, L. A. P. Zayas, J. Sonnenschein and D. Vaman, JHEP **0506**, 046 (2005) [hep-th/0410035]; S. Kuperstein and J. Sonnenschein, JHEP **0411**, 026 (2004) [hep-th/0411009].

17. I. Kirsch and D. Vaman, Phys. Rev. D **72**, 026007 (2005) [hep-th/0505164].

18. A. Armoni, M. Shifman and G. Veneziano, Phys. Rev. Lett. **91**, 191601 (2003) [hep-th/0307097]; Phys. Lett. B **579**, 384 (2004) [hep-th/0309013].

19. G. 't Hooft, Nucl. Phys. B **72**, 461 (1974).

20. L. Y. Glozman, Phys. Lett. B **539**, 257 (2002) [hep-ph/0205072]; T. D. Cohen and L. Y. Glozman, Int. J. Mod. Phys. A **17**, 1327 (2002) [hep-ph/0201242];

21. L. Y. Glozman, AIP Conf. Proc. **717**, 726 (2004) [hep-ph/0309334].

22. L. Y. Glozman, Phys. Lett. B **587**, 69 (2004) [hep-ph/0312354].

23. L. Y. Glozman, *Restoration of chiral symmetry in excited hadrons*, Lectures at 44-th Cracow School of Theoretical Physics *New Results in Particle Physics*, Zakopane, Poland, May 2004, hep-ph/0410194; this is a review where the reader can find references to earlier works.

24. V. A. Novikov, M. A. Shifman, A. I. Vainshtein and V. I. Zakharov, Nucl. Phys. B **191**, 301 (1981); M. A. Shifman, Sov. J. Nucl. Phys. **36**, 749 (1982) [Yad. Fiz. **36**, 1290 (1982)].

25. L. Y. Glozman, *Chiral and $U(1)_A$ restorations high in the hadron spectrum, semiclassical approximation and large N_c*, hep-ph/0411281.

26. A review can be found e.g. in T. D. Cohen and X. D. Ji, Phys. Rev. D **55**, 6870 (1997) [hep-ph/9612302]. For a discussion of SU(2)×SU(2) representations for baryons unrelated to dynamical issues relevant to high excitations the reader is referred to D. Jido, T. Hatsuda and T. Kunihiro, Phys. Rev. Lett. **84**, 3252 (2000) [hep-ph/9910375]; D. Jido, M. Oka and A. Hosaka, Prog. Theor. Phys. **106**, 873 (2001) [hep-ph/0110005].

27. M. V. Chizhov, *Tensor excitations in Nambu–Jona-Lasinio model*, hep-ph/9610220; JETP Lett. **80**, 73 (2004) [hep-ph/0307100].

28. G. 't Hooft, *Nucl. Phys.* **B75** (1974) 461 [Reprinted in G. 't Hooft, *Under the Spell of the Gauge Principle* (World Scientific, Singapore 1994), page 443]; see also F. Lenz, M. Thies, S. Levit and K. Yazaki, *Ann. Phys.* (N.Y.) **208** (1991) 1; C. Callan, N. Coote, and D. Gross, *Phys. Rev.* **D13** (1976) 1649; M. Einhorn, *Phys. Rev.* **D14** (1976) 3451; M. Einhorn, S. Nussinov, and E. Rabinovici, *Phys. Rev.* **D15** (1977) 2282.

29. I. Bars and M. Green, *Phys. Rev.* **D17** (1978) 537.

30. A. Casher, H. Neuberger and S. Nussinov, Phys. Rev. D **20**, 179 (1979).

31. B. Blok, M. A. Shifman and D. X. Zhang, Phys. Rev. D **57**, 2691 (1998); (E) D **59**, 019901 (1999) [hep-ph/9709333].

32. E. Schreiber, *Excited mesons and quantization of string endpoints*, hep-th/0403226.

33. M. A. Shifman, A. I. Vainshtein and V. I. Zakharov, Nucl. Phys. B **147**, 385; 448 (1979).

34. S. R. Beane, Phys. Rev. D **64**, 116010 (2001) [hep-ph/0106022].

35. S. S. Afonin, A. A. Andrianov, V. A. Andrianov and D. Espriu, JHEP **0404**, 039 (2004) [hep-ph/0403268].

36. J. M. Maldacena, Adv. Theor. Math. Phys. **2**, 231 (1998) [Int. J. Theor. Phys. **38**, 1113 (1999)] [hep-th/9711200]; S. S. Gubser, I. R. Klebanov and A. M. Polyakov, Phys. Lett. B **428**, 105 (1998) [hep-th/9802109]; E. Witten, Adv. Theor. Math. Phys. **2**, 253 (1998) [hep-th/9802150].

37. J.J. Sakurai, *Currents and Mesons*, (Univ. of Chicago Press, 1969).

38. J. Hirn and V. Sanz, *Interpolating between low and high energy QCD via a 5D Yang–Mills model*, hep-ph/0507049.

39. V. L. Chernyak and A. R. Zhitnitsky, *Asymptotic Behavior Of Exclusive Processes In QCD,* Phys. Rept. **112**, 173 (1984); S. J. Brodsky and G. P. Lepage, *Exclusive Processes In Quantum Chromodynamics,* Adv. Ser. Direct. High Energy Phys. **5**, 93 (1989) [reprinted in *Perturbative Quantum Chromodynamics,* Ed. A.H. Mueller (World Scientific, Singapore, 1989), p. 93].

40. A. A. Migdal, Annals Phys. **110**, 46 (1978); H. G. Dosch, J. Kripfganz and M. G. Schmidt, Phys. Lett. B **70**, 337 (1977).

HIDDEN QCD SCALES AND DIQUARK CORRELATIONS

ARKADY VAINSHTEIN

William I. Fine Theoretical Physics Institute,
University of Minnesota, Minneapolis, MN 55455

Numerically large QCD scales were discussed first in connection with a number of phenomena mostly related to vacuum quantum numbers and $0^{\pm}$ glueball and quark channels. We present arguments regarding the possible role of the larger scale in diquarks important for low-energy hadron phenomenology. Good diquarks, i.e. the 0^+ states of two quarks, are argued to have a two-component structure with one of the components peaking at distances several times shorter than a typical hadron size (a short-range core).

1. Introduction

The presentation is based on the work [1] with Misha Shifman where we study an impact of hierarchical scales in QCD on diquark features and related phenomena.

Let us start with the observation that the constituent quark model enjoy a remarkable success in qualitative and semiquantitative descriptions of a huge body of data on traditional hadron spectroscopy, static parameters and other regularities At the same time, since the advent of QCD it has been known that some important hadronic phenomena cannot be easily understood in this model. Probably, the most clear-cut example of this type is the masslessness of pion (in the chiral limit), which can only occur due to a "superstrong" attraction in the 0^- quark-antiquark channel. In the 1980's, a similar strong attraction was argued to exist in the 0^+ quark-quark channel. This conjecture was supported by occasional observations of a special role played by diquarks in low-energy hadronic phenomenology.

It was argued [2,3] that an instanton-induced interaction is sufficient to form a bound antitriplet scalar diquark. Then, there were suggestions [4] (see also the earlier analysis [5] and the recent discussion in [3]) that diquark correlations are important in the enhancement of $\Delta I = 1/2$ amplitudes of strangeness-changing decays.

Recently, the issue was revitalized by experimental evidence for existence of exotic pentaquark baryon, the state which was predicted to be narrow in Ref. [6] based on the chiral-quark-soliton model. Diquarks were suggested as an alternative explanation for pentaquark in a number of papers [7–10] which also discussed other phenomenological evidences. Although the existence of pentaquark is now strongly doubtful, implications of diquark correlations could be instrumental in excited and exotic hadron spectroscopy and in other aspects of hadronic physics. "Good" diquarks in color-flavor locked antisymmetric combination, due to a strong attraction, are probably bound to the extent that the "mass" of a good diquark roughly coincides with that of a constituent quark.

It is natural to suggest that the characteristic size of the good diquark as well as that of the constituent quark is considerably smaller than the nucleon size. In the case of the constituent quark a relevant momentum scale in the $\bar{q}q$ channel is of order of a few GeV. We will argue that a similar scale appears in the diquark qq channel. A basic tool here is the observation that if $SU(3)_{\text{color}}$ is replaced by $SU(2)_{\text{color}}$, diquarks become well-defined gauge-invariant objects, related by symmetry to conventional Goldstone bosons (pions).

The existence of hierarchical scales in QCD was discussed long ago in [11]. The largest of these scales shows up in $0^{\pm}$ glueball channels. QCD with massless flavors (the chiral limit) classically contains no scale and conformally invariant. Quantum effects break this invariance, and the physical scale Λ_{QCD} appears due to so called dimensional transmutation,

$$\Lambda_{\text{QCD}} = M_{\text{UV}} e^{-\frac{8\pi^2}{bg_{\text{UV}}^2}}.$$ (1)

It is in one-loop approximation for the running coupling g, M_{UV} is the ultraviolet cut-off, $b = \frac{11}{3}N_c - \frac{2}{3}N_f$ and $g_{\text{UV}} = g(M_{\text{UV}})$.

Phenomenologically Λ_{QCD} is of order of few hundred MeV. In terms of hadronic masses the one of the meson can be taken as a characteristic hadronic scale,

$$m_\rho = \text{const} \times \Lambda_{\text{QCD}} = 770 \text{ MeV}.$$ (2)

Although parametrically Λ_{QCD} is the only scale in QCD indeed there are some evidences for an existence of *numerically* much larger hadronic scales.

- $\Delta I = 1/2$ enhancement in weak nonleptonic decays, the $\Delta I = 1/2$ amplitudes are about 10 times larger than the $\Delta I = 3/2$ ones. Related is an enhancement in the ratio ϵ'/ϵ of the parameters of CP violation in K decays.

- A number of phenomena in heavy flavor decays, including a value of the Λ_b lifetime which is 10-15 % shorter than expected, large rates for $B \to \eta' X$ decays, importance of penguins in $b \to s$ transitions and problem with quark-hadron duality in $b \to u$ transitions.
- Large mass (958 MeV) of η'. Theoretically $m_{\eta'}^2/m_N^2 \propto 1/N_c^3$. In reality this ratio is close to 1.
- Enhancement in comparison to perturbative QCD estimates in many exclusive processes with large momentum transfer.
- Large duality interval in 0^-, 0^+ glueball and quark channels.
- Large s-wave $\pi\pi$ interaction without prominent resonances.
- A small value of α', a slope of the pomeron, the vacuum Regge trajectory.
- Jet production in e^+e^- annihilation – the total energy should be around 10 GeV to see clear-cut jets.
- Striking (and unexpected) experimental evidence from heavy ion collisions at RHIC. Instead of weakly coupled quark-gluon plasma experimentalists see a hydrodynamical flow.

2. Glueballs and large momentum scale

Breaking of conformal invariance in QCD by quantum effect can be expressed as an anomaly in the trace of energy-momentum tensor θ_μ^μ,

$$\theta_\mu^\mu = \frac{\beta(\alpha_s)}{4\alpha_s} G_{\mu\nu}^a G^{a\,\mu\nu}, \qquad \beta(\alpha_s)\mu\,\frac{d\alpha_s(\mu)}{d\mu} = -b\,\frac{\alpha_s}{2\pi} + \ldots \qquad (3)$$

The dilation feature of θ_μ^μ leads to the low-energy theorem [11],

$$\int dx \left\langle 0|T\left\{\frac{\alpha_s}{\pi} G_{\mu\nu}^2(x), \frac{\alpha_s}{\pi} G_{\gamma\delta}^2(0)\right\}|0\right\rangle = \frac{32}{b}\left\langle 0\left|\frac{\alpha_s}{\pi} G_{\mu\nu}^2(0)\right|0\right\rangle. \qquad (4)$$

Numerically, using the phenomenological value for gluon condensate [12]

$$\left\langle 0\left|\frac{\alpha_s}{\pi} G_{\mu\nu}^2(0)\right|0\right\rangle = (330\ \text{MeV})^4, \qquad (5)$$

we find a large momentum scale λ_G at which perturbative and nonperturbative effects are comparable,

$$\lambda_G^2 = 20\ \text{GeV}^2. \qquad (6)$$

Compare with the scale in the 1^- isovector quark channel

$$\lambda_\rho^2 = m_\rho^2 = 0.6\ \text{GeV}^2. \qquad (7)$$

The order of magnitude difference is quite significant.

3. Two-component structure of pion

We know about "abnormally strong" interaction in channel where the light pion exists. One can present a strong argument for a short-range core in the pion. Compare two matrix elements,

$$\langle 0|\bar{d}\gamma_\mu\gamma_5 u|\pi^+\rangle = if_\pi q_\mu ,$$

$$\langle 0|\bar{d}\gamma_5 u|\pi^+\rangle = if_\pi \frac{m_\pi^2}{m_u + m_d} . \tag{8}$$

The second matrix element is enhanced by the ratio

$$\frac{m_\pi^2}{m_u + m_d} \approx 1.8 \text{ GeV} . \tag{9}$$

It has a smooth chiral limit $m_q \to 0$ but is rather large numerically. Within QCD sum rules [12] it shows as difference in duality intervals,

$$\bar{d}\gamma_\mu\gamma_5 u : \qquad 0.6 \text{ GeV}^2,$$

$$\bar{d}\gamma_5 u : \qquad 1.9 \text{ GeV}^2. \tag{10}$$

How can one interpret this fact? In the quantum-mechanical approach the matrix elements (8) can be viewed as a value of wave function at zero separation, $\psi(0)$. We have two of them: the leading twist, $t = 2$ for $\bar{d}\gamma_\mu\gamma_5 u$ and nonleading twist, $t = 3$, for $\bar{d}\gamma_5 u$,

$$\psi = \alpha_2\psi_{t=2} + \alpha_3\psi_{t=3} . \tag{11}$$

While the coefficient α_3 is small the large value of $\psi_{t=3}(0)$ for the $\bar{d}\gamma_5 u$ component implies the existence of a smaller size.

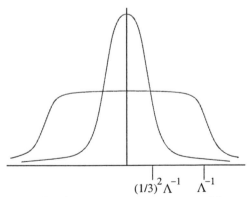

$$(1/3)^2\Lambda^{-1} \qquad \Lambda^{-1}$$

Figure 1. Double-component structure with a core.

We try to illustrate this in Fig. 1. The factor $(1/3)^2$ for core size can be interpreted in the framework of the instanton liquid model [13]. The model operates with two parameters, the average instanton size $\rho = 0.48\,\Lambda_{\mathrm{QCD}}^{-1}$ which is a factor of ~ 3 smaller than the average instanton separation $R = 1.35\,\Lambda_{\mathrm{QCD}}^{-1}$. Instantons are Euclidean objects and to relate their parameters with our Minkowski world, note that $R^{-1} \sim \Lambda_{\mathrm{QCD}}$ is a typical hadronic scale while ρ^{-1} is the geometrical mean between Λ_{QCD} and the higher glueball scale Λ_{gl}. Hence, we conclude that in the glueball world $\Lambda_{\mathrm{gl}} \sim 3^2\Lambda_{\mathrm{QCD}}$, which explains $(1/3)^2$ in Fig. 1.

4. Diquarks' progenitors

What about diquarks sizes? Can be related to the pion ones when the color SU(3) is replaced by SU(2). Flavor group becomes SU(4),

$$\chi^i = \left(u_L,\, d_L\,, \bar{d}_R\,, -\bar{u}_R\right) \tag{12}$$

and the the pattern of the spontaneous breaking is [14, 15].

$$\mathrm{SU}(4) \to \mathrm{SO}(5)\,. \tag{13}$$

A nonvanishing vacuum average of the six bilinears $\chi^i\chi^j = -\chi^j\chi^i$ can be aligned as

$$\langle \bar{u}u + \bar{d}d \rangle = \langle \bar{u}_R u_L + \bar{u}_L u_R + \bar{d}_R d_L + \bar{d}_L d_R \rangle \neq 0 \tag{14}$$

Acting on $\bar{u}u + \bar{d}d$ by SU(4) generators we determine five Goldstone bosons corresponding to operators

$$\bar{u}_R u_L - \bar{d}_R d_L - (R \leftrightarrow L) = -\bar{u}\gamma_5 u + \bar{d}\gamma_5 d\,,$$

$$\bar{u}_R d_L - (R \leftrightarrow L) = -\bar{u}\gamma_5 d\,, \qquad \bar{d}_R u_L - (R \leftrightarrow L) = -\bar{d}\gamma_5 u\,, \tag{15a}$$

$$u_L d_L + (R \leftrightarrow L) = -u C\gamma_5 d\,, \qquad \bar{d}_L \bar{u}_L + (R \leftrightarrow L) = -\bar{d}C\gamma_5\bar{u}\,. \tag{15b}$$

In addition to the triplet of massless 0^- pions (first two lines) we get two "baryon" Goldstone bosons (the third line) which are 0^+ diquark states.

Thus, we have more than the conventional triplet of pions; there emerge two extra diquark states related to pions by exact symmetry. Since in SU(2)$_{\mathrm{color}}$ theory baryonic Goldstones and pions are related by an exact symmetry, their spatial structure is the same. In particular, the two-component structure depicted in Fig. 1 is shared by both. As we pass to SU(3)$_{\mathrm{color}}$, the exact symmetry no longer holds, but an approximate similarity in the spatial structure of pions and diquarks is expected to hold.

5. Conclusions

There is a number of possible applications which are not discussed in this note. In particular, see [1], the core-like structure in diquarks can play a role in solving two old puzzles of the 't Hooft $1/N$ expansion: strong quark-mass dependence of vacuum energy density and strong violations of the Okubo–Zweig–Iizuka rule in quark-antiquark $0^{\pm}$ channels. In both cases empiric data defy 't Hooft's $1/N$ suppression.

This work was supported in part by DOE grant DE-FG02-94ER408.

References

1. M. Shifman and A. Vainshtein, Phys. Rev. D **71**, 074010 (2005).
2. T. Schäfer, E. V. Shuryak and J. J. M. Verbaarschot, Nucl. Phys. B **412**, 143 (1994).
3. M. Cristoforetti, P. Faccioli, E. V. Shuryak and M. Traini, Phys. Rev. D **70**, 054016 (2004).
4. B. Stech, Phys. Rev. D **36**, 975 (1987); H. G. Dosch, M. Jamin and B. Stech, Z. Phys. C **42**, 167 (1989); B. Stech and Q. P. Xu, Z. Phys. C **49**, 491 (1991); M. Neubert and B. Stech, Phys. Rev. D **44**, 775 (1991).
5. M. A. Shifman, A. I. Vainshtein and V. I. Zakharov, Sov. Phys. JETP **45**, 670 (1977).
6. D. Diakonov, V. Petrov and M. V. Polyakov, Z. Phys. A **359**, 305 (1997).
7. M. Karliner and H. J. Lipkin, hep-ph/0307243; hep-ph/0307343; Phys. Lett. B **575**, 249 (2003); **612**, 197 (2005).
8. R. L. Jaffe and F. Wilczek, Phys. Rev. Lett. **91**, 232003 (2003); Phys. Rev. D **69**, 114017 (2004); Phys. World **17**, 25 (2004).
9. E. Shuryak and I. Zahed, Phys. Lett. B **589**, 21 (2004).
10. R. L. Jaffe, *Exotica*, Phys. Rep. **409**, 1 301 (2005).
11. V. A. Novikov, M. A. Shifman, A. I. Vainshtein and V. I. Zakharov, Nucl. Phys. B **191**, 301 (1981).
12. M. A. Shifman, A. I. Vainshtein and V. I. Zakharov, Nucl. Phys. B **147**, 385; 448 (1979).
13. E. V. Shuryak, Nucl. Phys. B **203**, 93, 116, 140 (1982); **302**, 559, 574, 599, 621 (1988); D. Diakonov and V. Y. Petrov, Nucl. Phys. B **272**, 457 (1986); **245**, 259 (1984);
14. S. Dimopoulos, Nucl. Phys. B **168**, 69 (1980); M. E. Peskin, Nucl. Phys. B **175**, 197 (1980).
15. I. Kogan, M. Shifman and M. Vysotsky, Sov. J. Nucl. Phys. **42**, 318 (1985) [Yad. Fiz. **42**, 504 (1985)].

NEARLY CONFORMAL QCD AND ADS/CFT *

GUY F. DE TÉRAMOND

Universidad de Costa Rica, San José, Costa Rica

STANLEY J. BRODSKY

Stanford Linear Accelerator Center
Stanford University, Stanford, CA 94309, USA

The AdS/CFT correspondence is a powerful tool to study the properties of conformal QCD at strong coupling in terms of a higher dimensional dual gravity theory. The power-law falloff of scattering amplitudes in the non-perturbative regime and calculable hadron spectra follow from holographic models dual to QCD with conformal behavior at short distances and confinement at large distances. String modes and fluctuations about the AdS background are identified with QCD degrees of freedom and orbital excitations at the AdS boundary limit. A description of form factors in space and time-like regions and the behavior of light-front wave functions can also be understood in terms of a dual gravity description in the interior of AdS.

The correspondence[1], between string theory in a warped 10-dimensional space and conformal Yang-Mills theories defined at its four dimensional space-time boundary at infinity has led to important insights into the properties of QCD at strong coupling. The applications include the nonperturbative derivation[2] of dimensional counting rules[3] for hard exclusive glueball scattering, the description of deep inelastic structure functions[4] and the power falloff of hadronic light-front wave functions (LFWF) including orbital angular momentum[5]. In the original correspondence[1], the low energy supergravity approximation to type IIB string compactified on $AdS_5 \times S^5$, is dual to the $\mathcal{N} = 4$ super Yang-Mills (SYM) theory at large N_C.

QCD is fundamentally different from SYM theories where all of the matter fields transform in adjoint multiplets of $SU(N_C)$. Its string dual is unknown. We assume however that such a string exists and that, in principle, it can be defined in terms of the QCD degrees of freedom at infinity. In practice, we can deduce some of the dual string properties by

*This work is supported by the Department of Energy contract DE–AC02–76SF00515.

studying its boundary ultraviolet limit, $r \to \infty$, as well as the small-r infrared region of AdS space, characteristic of strings dual to confining gauge theories. This approach, which can be described as a bottom-up approach, has been successful in obtaining general properties of the low-lying hadron spectra, chiral symmetry breaking and hadron couplings in AdS/QCD[6], in addition to the scattering results described in[2,4,5]. A different approach, a top-bottom approach, consists in studying the full supergravity equations to compute the glueball spectrum[7]. The addition of D3/D7 branes[8] or the gravitational fluctuations of a thick brane in Minkowski space[9] leads to a calculable meson spectrum. Other aspects of high energy scattering in warped spaces have been addressed in[10].

It is remarkable that dimensional scaling for exclusive processes works so well at relatively low energies where higher twist effects are expected to be dominant[11]. Counting rules can be understood if QCD resembles a strongly coupled theory at moderate energies. The isomorphism of the group $SO(2,4)$, which act as the group of conformal symmetries at the AdS boundary in the limit of massless quarks and vanishing β function, with the group of isometries of AdS, $x^\mu \to \lambda x^\mu$, $r \to r/\lambda$, maps scale transformations into the holographic coordinate r. Consequently, the string mode in r is the extension of the hadron wave function into the fifth dimension. In particular, the $r \to 0$ boundary corresponds to the zero separation limit between quarks. Conversely, color confinement implies that there is a maximum separation of quarks and a minimum value $r_0 = \Lambda_{QCD} R^2$, where the string modes can propagate. The cutoff at r_0 is dual to the introduction of a mass gap Λ_{QCD}, it breaks the conformal invariance and is responsible for the generation of a spectrum of color singlet hadronic states.

The duality between a gravity theory on AdS_{d+1} space and the strong coupling limit of a conformal gauge theory at its d-dimensional boundary, is given in terms of the $d+1$ partition function in the bulk $Z_{grav}[\Phi(x,z)] = \int \mathcal{D}[\Phi]e^{iS_{grav}[\Phi]}$, and the d-dimensional functional integral over quarks q and gluons G in presence of an external source,

$$Z_{QCD}[\Phi_o(x)] = \int [DG][Dq] \exp\left\{ iS_{QCD}[G,q] + i \int d^d x \Phi_o \mathcal{O} \right\}, \quad (1)$$

with conditions[12] $Z_{grav}\left[\Phi(x,z)_{z=0} = \Phi_o(x)\right] = Z_{QCD}\left[\Phi_o\right]$, where $z = R^2/r$ and R is the AdS radius. Near the AdS boundary, $z \to 0$, $\Phi(x,z)$ behaves as $\Phi(x,z) \to z^\Delta \Phi_+(x) + z^{d-\Delta} \Phi_-(x)$, where $\Phi_-(x)$ is the boundary source, $\Phi_- = \Phi_o$, and $\Phi_+(x)$ is the normalizable solution with conformal dimension Δ. The physical string modes $\Phi(x,z) \sim e^{-iP \cdot x} f(r)$, are plane waves along the Poincaré coordinates with four-momentum P^μ and hadronic invariant mass states $P_\mu P^\mu = \mathcal{M}^2$. For large-$r$, $f(r) \sim r^{-\Delta}$. The dimension Δ of the string mode, is the same dimension of the interpolating operator $\mathcal{O}$ which creates a hadron out of the vacuum: $\langle P|\mathcal{O}|0\rangle \neq 0$.

QCD degrees of freedom are defined at the AdS boundary at infinity. Quarks and gluons also propagate in the AdS interior. In the limit where the linearized equations for spin 0, $\frac{1}{2}$, 1 and $\frac{3}{2}$ on $AdS_5 \times S^5$ have no interactions, only color singlet states of dimension 3, 4 and $\frac{9}{2}$ have dual string modes and a physical spectrum. Consequently, only the hadronic states (dimension-3) $J^P = 0^-, 1^-$ pseudoscalar and vector mesons, the (dimension-$\frac{9}{2}$) $J^P = \frac{1}{2}^+, \frac{3}{2}^+$ baryons, and the (dimension-4) $J^P = 0^+$ glueball states, corresponding exactly to the lowest-mass physical hadronic states can be derived in the classical holographic limit[13]. Hadrons fluctuate in particle number, but there are also orbital angular momentum fluctuations. A major difficulty in describing the hadron spectrum with AdS/CFT arises from the nature of the string solutions, since duality cannot be established for spin > 2, where the conformal dimensions become very large. Higher Fock components are manifestations of the quantum fluctuations of QCD; metric fluctuations of the bulk geometry about the fixed AdS background should correspond to quantum fluctuations of Fock states above the valence state. Indeed, for large Lorentz spin, orbital excitations in the boundary correspond to quantum fluctuations about the AdS metric[14]. We identify the higher spin hadrons with the fluctuations around the spin 0, $\frac{1}{2}$, 1 and $\frac{3}{2}$ classical string solutions on the AdS_5 sector[13].

As a specific example, consider the twist (dimension minus spin) two glueball interpolating operator $\mathcal{O}_{4+L}^{\ell_1\cdots\ell_m} = FD_{\{\ell_1}\ldots D_{\ell_m\}}F$ with total internal space-time orbital momentum $L = \sum_{i=1}^m \ell_i$ and conformal dimension $\Delta_L = 4 + L$. We match the large r asymptotic behavior of each string mode to the corresponding conformal dimension of the boundary operators of each hadronic state while maintaining conformal invariance. In the conformal limit, an L quantum, which is identified with a quantum fluctuation about the AdS geometry, corresponds to an effective five-dimensional mass μ in the bulk side. The allowed values of μ are uniquely determined by requiring that asymptotically the dimensions become spaced by integers, according to the spectral relation $(\mu R)^2 = \Delta_L(\Delta_L - 4)$[13]. The interaction term in Eq. 1 for a state with orbital L at the asymptotic boundary results in the effective coupling

$$S_{int} = \int d^4x \, \partial_{x_{\ell_1}} \cdots \partial_{x_{\ell_m}} \Phi(x,z)|_{z=0} \, \mathcal{O}^{\ell_1\cdots\ell_m}. \tag{2}$$

The string modes $\Phi^{\ell_1 \cdots \ell_m}(x, z) = \partial_{x_{\ell_1}} \cdots \partial_{x_{\ell_m}} \Phi(x, z)$ for a given eigenvalue μR, with $\Phi(x, z) = C e^{-iP \cdot x} z^2 J_\alpha(z \beta_{\alpha,k} \Lambda_{QCD})$, $\alpha = \Delta_L - 2$, have the correct Lorentz structure at the AdS boundary, since each ∂_{x^μ} pulls down a P_μ from the exponential factor of the string mode, leaving intact the holographic z-dependence which is determined by the conformal dimension Δ_L.

The four-dimensional mass spectrum follows from the Dirichlet boundary condition $\Phi(x, z_o) = 0$, $z_0 = 1/\Lambda_{QCD}$, on the AdS string modes for the different wave functions corresponding to spin < 2 and is given in terms of the zeros of Bessel functions, $\beta_{\alpha,k}$. In the case of mesons the predicted spectrum is shown in Figure 1 for $\Lambda_{QCD} = 0.263$ GeV, the only parameter in the model. The baryon spectrum is discussed in Ref. 13.

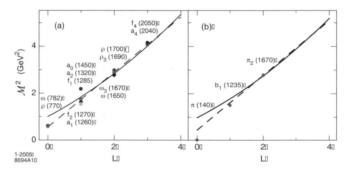

Figure 1. Light meson orbital states for $\Lambda_{QCD} = 0.263$ GeV: (a) vector mesons and (b) pseudoscalar mesons. The dashed line is a linear Regge trajectory with slope 1.16 GeV^2.

The form factor in AdS/QCD is the overlap of the normalizable modes dual to the incoming and outgoing hadron Φ_I and Φ_F and the non-normalizable mode $J(Q, z)$, dual to the external source

$$F(Q^2)_{I \to F} \simeq R^{3+2\sigma} \int_0^{z_o} \frac{dz}{z^{3+2\sigma}} \, \Phi_F(z) \, J(Q, z) \, \Phi_I(z), \tag{3}$$

where $\sigma_n = \sum_{i=1}^{n} \sigma_i$ is the spin of the interpolating operator $\mathcal{O}_n$, which creates an n-Fock state $|n\rangle$ at the AdS boundary. $J(Q, z)$ has the value 1 at zero momentum transfer, and as boundary limit the external current, thus $A^\mu(x, z) = \epsilon^\mu e^{iQ \cdot x} J(Q, z)$. The solution to the AdS wave equation subject to boundary conditions at $Q = 0$ and $z \to 0$ is[4] $J(Q, z) = zQ K_1(zQ)$. At large enough $Q \sim r/R^2$, the important contribution to (3) is from the region near $z \sim 1/Q$. At small z, the n-mode $\Phi^{(n)}$ scales as $\Phi^{(n)} \sim z^{\Delta_n}$, and we recover the power law scaling[3], $F(Q^2) \to [1/Q^2]^{\tau_n - 1}$, where the twist $\tau_n = \Delta_n - \sigma_n$, is equal to the number of partons, $\tau_n = n$. A numerical computation for the pion form factor[15] gives the results shown in Figure 2, where the resonant structure in the time-like region from the AdS cavity modes is apparent.

The AdS/QCD correspondence provides also a simple description of hadrons at the amplitude level by mapping string modes to the impact space representation of LFWF. In terms of partonic variables $x_i \vec{r}_{\perp i} = \vec{R}_\perp + \vec{b}_{\perp i}$, where $\vec{r}_{\perp i}$ are the physical transverse position coordinates, $\vec{b}_{\perp i}$ internal coordinates, $\sum_i \vec{b}_{\perp i} = 0$, and $\vec{R}_\perp$ the hadron transverse center of momentum $\vec{R}_\perp = \sum_i x_i \vec{r}_{\perp i}$, $\sum_i x_i = 1$, we find for a two-parton LFWF

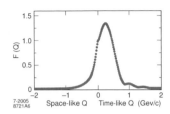

Figure 2. Space and time-like structure for the pion form factor in AdS/QCD.

$$\psi_{\ell,k}(x, \zeta) = C x (1-x) J_{1+\ell} \left(\zeta \beta_{1+\ell,k} \Lambda_{QCD} \right) / \zeta,$$

where $\zeta = |\vec{b}_\perp| \sqrt{x(1-x)}$ represents the scale of the invariant separation between quarks. The first eigenmodes are depicted in Figure 3.

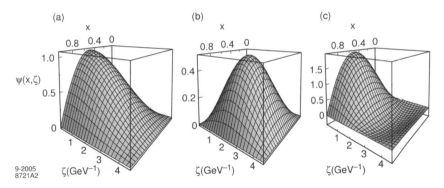

Figure 3. Two-parton bound state holographic LFWF $\psi(x, \zeta)$: (a) ground state $\ell = 0$, $k = 1$, (b) first orbital excited state $\ell = 1$, $k = 1$, (c) first radial exited state $\ell = 0$, $k = 2$.

The holographic model is quite successful in describing the known light hadron spectrum. The only mass scale is Λ_{QCD}. The model incorporates confinement and conformal symmetry. Only dimension 3, $\frac{9}{2}$ and 4 states $\bar{q}q$, qqq, and gg appear in the duality at the classical level. Non-zero orbital angular momentum and higher Fock-states require the introduction of quantum fluctuations. The model gives a simple description of the structure of hadronic form factors and LFWFs. It explains the suppression of the odderon. The dominance of quark-interchange in hard exclusive processes emerges naturally from the classical duality of the holographic model.

References

1. J. M. Maldacena, Adv. Theor. Math. Phys. **2**, 231 (1998).
2. J. Polchinski and M. J. Strassler, Phys. Rev. Lett. **88**, 031601 (2002).
3. S. J. Brodsky and G. R. Farrar, Phys. Rev. Lett. **31**, 1153 (1973); V. A. Matveev, R. M. Muradian and A. N. Tavkhelidze, Lett. Nuovo Cimento **7**, 719 (1973).
4. J. Polchinski and M. J. Strassler, JHEP **0305**, 012 (2003).
5. S. J. Brodsky and G. F. de Téramond, Phys. Lett. B **582**, 211 (2004).
6. H. Boschi-Filho and N. R. F. Braga, Eur. Phys. J. C **32**, 529 (2004); JHEP **0305**, 009 (2003); G. F. de Teramond and S. J. Brodsky, arXiv:hep-th/0409074; J. Erlich, E. Katz, D. T. Son and M. A. Stephanov, arXiv:hep-ph/0501128; S. Hong, S. Yoon and M. J. Strassler, arXiv:hep-ph/0501197; L. Da Rold and A. Pomarol, arXiv:hep-ph/0501218; H. Boschi-Filho, N. R. F. Braga and H. L. Carrion, arXiv:hep-th/0507063.
7. C. Csaki, H. Ooguri, Y. Oz and J. Terning, JHEP **9901**, 017 (1999); R. de Mello Koch, A. Jevicki, M. Mihailescu and J. P. Nunes, Phys. Rev. D **58**, 105009 (1998); C. Csaki, Y. Oz, J. Russo and J. Terning, Phys. Rev. D **59**, 065012 (1999); J. A. Minahan, JHEP **9901**, 020 (1999); R. C. Brower, S. D. Mathur and C. I. Tan, Nucl. Phys. B **587**, 249 (2000); E. Caceres and C. Nunez, arXiv:hep-th/0506051.
8. A. Karch, E. Katz and N. Weiner, Phys. Rev. Lett. **90**, 091601 (2003); M. Kruczenski, D. Mateos, R. C. Myers and D. J. Winters, JHEP **0307**, 049 (2003); JHEP **0405**, 041 (2004); T. Sakai and J. Sonnenschein, JHEP **0309**, 047 (2003); J. Babington, J. Erdmenger, N. J. Evans, Z. Guralnik and I. Kirsch, Fortsch. Phys. **52**, 578 (2004); C. Nunez, A. Paredes and A. V. Ramallo, JHEP **0312**, 024 (2003); S. Hong, S. Yoon and M. J. Strassler, JHEP **0404**, 046 (2004); arXiv:hep-th/0409118; M. Kruczenski, L. A. P. Zayas, J. Sonnenschein and D. Vaman, arXiv:hep-th/0410035; T. Sakai and S. Sugimoto, Prog. Theor. Phys. **113**, 843 (2005); arXiv:hep-th/0507073; A. Paredes and P. Talavera, Nucl. Phys. B **713**, 438 (2005); I. Kirsch and D. Vaman, arXiv:hep-th/0505164.
9. G. Siopsis, arXiv:hep-th/0503245.
10. S. B. Giddings, Phys. Rev. D **67**, 126001 (2003); O. Andreev and W. Siegel, Phys. Rev. D **71**, 086001 (2005); K. Kang and H. Nastase, arXiv:hep-th/0410173; H. Nastase, arXiv:hep-th/0501039; arXiv:hep-th/0501068.
11. Contributions of A. Mirazita, H. Gao and A. Deur to these proceedings.
12. S. S. Gubser, I. R. Klebanov and A. M. Polyakov, Phys. Lett. B **428**, 105 (1998); E. Witten, Adv. Theor. Math. Phys. **2**, 253 (1998).
13. G. F. de Teramond and S. J. Brodsky, Phys. Rev. Lett. **94**, 201601 (2005).
14. S. S. Gubser, I. R. Klebanov and A. M. Polyakov, Nucl. Phys. B **636**, 99 (2002).
15. S. J. Brodsky and G. F. de Téramond, in preparation.

MATCHING MESON RESONANCES TO OPE IN QCD*

S. S. AFONIN AND D. ESPRIU†

Departament d'Estructura i Constituents de la Matèria,
Universitat de Barcelona,
647 via Diagonal, Barcelona 08028, Spain
E-mail: afonin@ecm.ub.es, espriu@ecm.ub.es

A. A. ANDRIANOV

Istituto Nazionale di Fisica Nucleare, Sezione di Bologna,
via Irnerio 46, Bologna 40126, Italy
E-mail: alexandr.andrianov@bo.infn.it

V. A. ANDRIANOV

V.A. Fock Department of Theoretical Physics, St. Petersburg State University,
1 Ulyanovskaya ul., St. Petersburg 198905, Russia
E-mail: vladimir.andrianov@pobox.spbu.ru

We investigate for the light quark sector the possible corrections to the linear Regge trajectories by matching two-point correlators of quark currents to the Operator Product Expansion. We find that the allowed modifications to the linear behavior must decrease rapidly with the principal quantum number. After fitting the lightest states in each channel and certain low-energy constants the whole spectrum for meson masses and residues is obtained in a satisfactory agreement with phenomenology. The perturbative corrections to our results are discussed.

The observed masses squared of mesons with given quantum numbers form linear trajectories [1,2] depending on the number of radial excitation n. This is a strong indication that QCD admits an effective string description, as this type of spectrum is characteristic e.g. of the bosonic string. In the bosonic string model the slope of all trajectories must be equal since this

---

*This work is supported by RFBR, grant 05-02-17477, grant UR 02.01.299 and INFN, grant IS/PI13

†Work of S. A. and D. E. is partially supported by CYT FPA, grant 2004-04582-C02-01 and CIRIT GC, grant 2001SGR-00065.

quantity is proportional to the string tension depending on gluedynamics only. However, there exist sizeable deviations from the string picture which are not understood yet. In the present analysis we examine possible corrections to the linear trajectories in the vector (V), axial-vector (A), scalar (S), and pseudoscalar (P) channels [3]. Our method is based on the consideration of the two-point correlators of V,A,S,P quark currents in the large-N_c limit of QCD [4]. On the one hand, by virtue of confinement they are saturated by an infinite set of narrow meson resonances, that is, they can be represented by the sum of related meson poles in Euclidean space:

$$\Pi_J(Q^2) = \int d^4x \exp(iQx) \langle \bar{q}\Gamma q(x) \bar{q}\Gamma q(0) \rangle_{\text{planar}} = \sum_n \frac{2F_J^2(n)}{Q^2 + m_J^2(n)}, \quad (1)$$

$$J \equiv S, P, V, A; \qquad \Gamma = i, \gamma_5, \gamma_\mu, \gamma_\mu\gamma_5,$$

expressing the quark-hadron duality[5]. Further we denote $F_{S,P} \equiv G_{S,P}m_{S,P}$. On the other hand, their high-energy asymptotics is provided by the perturbation theory and the Operator Product Expansion (OPE) with condensates [6]. For instance, in the V,A channels:

$$\Pi_{V,A}(Q^2) = \frac{1}{8\pi^2}\left(\ln\frac{\Lambda^2}{Q^2} + \frac{4}{\beta_0}\ln\frac{\alpha_s(\Lambda^2)}{\alpha_s(Q^2)}\right) + \sum_{n=2}^{\infty}\frac{\langle O_{2n}^{V,A}\rangle}{Q^{2n}}. \quad (2)$$

Matching these two approaches results in certain Chiral Symmetry Restoration sum rules representing a set of constraints on meson mass parameters [7]. For simplicity we performed the analysis in the chiral limit and in the leading order of perturbation theory.

Let us consider the linear ansatz for the meson mass spectra with a non-linear correction δ:

$$m_J^2(n) = m_{0,J}^2 + a\,n + \delta_J(n), \qquad J \equiv V, A, S, P. \quad (3)$$

The last term in Eq. (3) signifies a possible deviation from the string picture in QCD. The condition of convergence for the generalized Weinberg sum rules [9] imposes the equality of slopes and a falloff of non-linear correction.

The asymptotic freedom leads to the relation between residues and masses: $F^2(n) \sim \frac{dm^2(n)}{dn}$. Our analysis showed that the analytical structure of the OPE admits, however, exponentially small deviations from this relation (compare to [9,10]). We can argue also that D-wave vector meson must asymptotically drop out the CSR sum rules. This results in the exponential decreasing of the corresponding residues. Our final results read:

$$m_J^2(n) = M_J^2 + an + A_m^J e^{-B_m \cdot n}, \tag{4}$$

$$F_{V,A}^2(n) = a \left(\frac{1}{8\pi^2} + A_F^{V,A} e^{-B_F \cdot n} \right), \tag{5}$$

$$G_{S,P}^2(n) = a \left(\frac{3}{16\pi^2} + A_G^{S,P} e^{-B_G \cdot n} \right) \tag{6}$$

with certain constants $A_{m,F,G}^{V,A,S,P}$ and $B_{m,F,G} > 0$ to be fitted. We do not know the underlying dynamics responsible for the appearance of those exponential corrections. But it seems for us reasonable to suppose that, for masses, this dynamics is governed mostly by gluons and thereby does not depend on flavor. Thus, we keep the exponent B_m the same for all channels. For the same reason we regard $B_{F,G}$ as independent of parity.

One has to decide which particles are chiral partners in order to guarantee chirally symmetric results. One may think of two possibilities: a) the π-meson belongs to the radial Regge trajectory being the parity-odd partner of the lightest scalar meson, b) it does not belong to the radial Regge trajectory being an isolated Goldstone chiral particle and therefore the lightest scalar meson is the parity-even partner of the $\pi'(1300)$ meson. In case a) the effective low-energy theory would be the linear σ-model, and in case b) it corresponds to the nonlinear σ-model. We checked the both variants (see below).

In order to compare the sum of the resonances with the OPE we have to choose appropriate values for the inputs of the latter. They are the condensates $\langle \bar{q}q \rangle$ and $\langle (G_{\mu\nu}^a)^2 \rangle$, and α_s. On the resonance side we take as input the pion decay constant f_π, the pion pole residue $Z_\pi = 2\frac{\langle \bar{q}q \rangle^2}{f_\pi^2}$ (from current algebra), and the slope a. The numerical values for all these parameters are taken at about the CSB scale ~ 1 GeV and presented in Appendix. In particular, the value of $f_\pi \approx 103$ MeV is certainly different from its low-energy limit [11] and corresponds to the matching of the OPE and the sums of resonances at the above scale.

Given an ansatz for mass spectrum $m_J^2(n)$ and decay constants F_J^2, one can calculate the electromagnetic pion mass difference $\Delta m_\pi \equiv m_{\pi^+} - m_{\pi^0}$ and the chiral constant L_{10} [11] (parameterizing the decay $\pi \to e\nu\gamma$). The example of such a calculation was demonstrated in [12]. Besides, in the scalar sector we can calculate the chiral constant L_8 [11] (parametrizing the ratio of current quark masses). The corresponding numerical results are presented in Table 1 of Appendix.

Let us discuss the impact of running coupling constant at next-to-leading order of perturbation theory. We have in this case the contribution to the imaginary part of correlator which is related to the full correlator through the dispersion relation. In the vector and axial-vector case this contribution is:

$$\mathrm{Im}\Pi(t) = \frac{1}{4\pi^2}\left(1 + \frac{\alpha_s(t)}{\pi}\right) + \mathcal{O}(\alpha_s^2). \qquad (7)$$

In the large-N_c limit the imaginary part is saturated by narrow meson states:

$$\mathrm{Im}\Pi(t) = \pi \sum_{n=0}^{\infty} 2F^2(n)\delta\left(t - m^2(n)\right). \qquad (8)$$

We perform summation in Eq. (8) by applying the Euler-Maclaurin summation formula. This provides smoothness of this expression. As a result:

$$\mathrm{Im}\Pi(t) \simeq \frac{2F^2(n_0)}{\left|\dfrac{dm^2(n)}{dn}\right|_{n=n_0}} \qquad (n_0 = n_0(t)). \qquad (9)$$

Comparing the result with Eq. (7) we obtain the perturbative correction for residues:

$$F^2(n_0) = \frac{dm^2(n)}{dn}\frac{1}{8\pi^2}\left(1 + \frac{\alpha_s\left(t(n_0)\right)}{\pi}\right). \qquad (10)$$

Eq. (10) can be approximated by finite differences. For the first two states this reads then:

$$\frac{F_\rho^2}{m_{\rho'}^2 - m_\rho^2} \approx \frac{F_{a_1}^2}{m_{a_1'}^2 - m_{a_1}^2} \approx \frac{1}{8\pi^2}\left(1 + \frac{\alpha_s(1\,\mathrm{GeV})}{\pi}\right). \qquad (11)$$

Substituting experimental values [13] into Eq. (11) one arrives at the estimate: $1.5\pm0.2 \approx 1.3\pm0.6 \approx 1.4$ in a good agreement with the phenomenology.

In order to reproduce the running coupling behaviour in Eq. (7) one should accept the following ansatz for the residues:

$$F^2(n) \simeq \frac{dm^2(n)}{dn}\frac{1}{8\pi^2}\left(1 + 4\left(\beta_0 \ln \frac{m^2(n)}{\Lambda_{\mathrm{QCD}}^2}\right)^{-1}\right) + \mathrm{exp.\ corr.} \qquad (12)$$

However this ansatz introduces into the OPE the additional terms $\sim \frac{\ln Q^2}{Q^2}$ (and powers of these terms) which are absent in the standart OPE. Thus ansatz (11) must be improved. This problem is under our studies now.

Let us summarize the results of our analysis:

1) The convergence of the generalized Weinberg sum rules requires the universality of slopes and intercepts for parity conjugated trajectories.

2) The matching to the OPE cannot be achieved by a simple linear parameterization of the mass spectrum, the linear trajectory ansatz. There must exist deviations from the linear trajectory ansatz triggered by chiral symmetry breaking. These deviations must decrease at least exponentially with n.

3) For heavy states, the D-wave vector mesons have to decouple from asymptotic sum rules. This fact implies the exponential (or faster) decreasing the corresponding decay constants $F_D^2(n)$.

4) Our results seem to exclude a light $\sigma(600)$ particle as a quarkonium state and rather favor the non-linear realization of chiral symmetry with the lightest scalar of mass ~ 1 GeV, its chiral partner being the $\pi'(1300)$.

5) In our approach the quantities L_8, L_{10} and Δm_π are obtained, in satisfactory agreement with the phenomenology.

6) Perturbative corrections can be systematically treated making, nevertheless, small effects on the fits presented.

Appendix A. Tables

In this Appendix we give an example of the meson mass spectra resulting from our work. The inputs general for all tables (if any) are: $a = (1120\,\text{MeV})^2$, $\langle \bar{q}q \rangle = -(240\,\text{MeV})^3$, $\frac{\alpha_s}{\pi}\langle(G^a_{\mu\nu})^2\rangle = (360\,\text{MeV})^4$, $f_\pi = 103\,\text{MeV}$, $Z_\pi = 2\frac{\langle \bar{q}q \rangle^2}{f_\pi^2}$, $\alpha_s = 0.3$. The units are: $m(n), F(n), G(n)$ — MeV; A_m — MeV^2; A_F, A_G, $B_{F,G,m}$ — MeV^0.

Table 1. An example of parameters for the mass spectra of our work. The corresponding experimental values [2,12] (if any) are displayed in brackets.

Case	Inputs	Fits and constants
VA	$m_V(0) = 770\,(769.3 \pm 0.8)$, $m_A(0) = 1200\,(1230 \pm 40)$,	$M = 920$, $B_m = 0.97$, $B_F = 0.72$, $A_m^V = -500^2$, $A_m^A = 770^2$, $A_F^V = 0.0012$, $A_F^A = -0.0031$, $L_{10} = -6.5 \cdot 10^{-3}$, $\Delta m_\pi = 2.3$
SP (π-in)	$m_S(0) = 1000$, $m_P(0) = 0$, $m_P(1) = 1300\,(1300 \pm 100)$, $B_m = 0.97$	$M = 840$, $B_G = 0.42$, $A_m^S = 550^2$, $A_m^P = -840^2$, $A_G^S = -0.0009$, $A_G^P = 0.0004$, $L_8 = 1.0 \cdot 10^{-3}$
SP (π-out)	$m_S(0) = 1000$, $m_P(0) = 1300\,(1300 \pm 100)$, $m_P(1) = 1800\,(1801 \pm 13)$, $B_m = 0.97$	$M = 1470$, $B_G = 1.27$, $A_m^S = -1080^2$, $A_m^P = -690^2$, $A_G^S = 0.0213$, $A_G^P = 0.0067$, $L_8 = 0.9 \cdot 10^{-3}$

Table 2. Mass spectrum and residues for the parameter sets of Table 1.

n	out	0	1	2	3
$m_V(n)$		770 (775.8 ± 0.5)	1420 (1465 ± 25)	1820 (1900?)	2140 (2149 ± 17)
$F_V(n)$		138 (154±8)	135	133	133
$m_A(n)$		1200 (1230 ± 40)	1520 (1640 ± 40)	1850 (1971 ± 15)	2150 (2270 ± 50)
$F_A(n)$		116 (123 ± 25)	125	128	130
$m_S(n)$		1000 (980 ± 10)	1440 (1507 ± 5)	1800 (1714 ± 5)	2100
$G_S(n)$		176	178	178	179
$m_P(n)$		0 (π-in)	1300 (1300 ± 100)	1760 (1801 ± 13)	2100 (2070 ± 35)
$G_P(n)$		–	179	179	179
$m_S(n)$		1000 (980 ± 10)	1730 (1714 ± 5)	2120	2420
$G_S(n)$		243	199	185	181
$m_P(n)$	0	1300 (1300 ± 100)	1800 (1812 ± 14)	2150 (2070 ± 35)	2430 (2360 ± 30)
$G_P(n)$	–	201	186	181	180

References

1. A. Bramon, E. Etim and M. Greco, *Phys. Lett.* **B41**, 609 (1972); M. Greco, *Nucl. Phys.* **B63**, 398 (1973); J. J. Sakurai, *Phys. Lett.* **B46**, 207 (1973).
2. A. V. Anisovich, V. V. Anisovich and A. V. Sarantsev, *Phys. Rev.* **D62**, 051502 (2000).
3. S. S. Afonin, A. A. Andrianov, V. A. Andrianov and D. Espriu, *JHEP* **039** (2004).
4. G. 't Hooft, *Nucl. Phys.* **B72**, 461 (1974); E. Witten, *Nucl. Phys.* **B160**, 57 (1979).
5. M. Shifman, *At the Frontier of Particle Physics/Handbook of QCD*, (ed. M. Shifman), World Scientific, 2001; hep-ph0009131; see also hep-ph/0507246.
6. M. A. Shifman, A. I. Vainstein and V. I. Zakharov, *Nucl. Phys.* **B147**, 385, 448 (1979).
7. A. A. Andrianov and V. A. Andrianov, hep-ph/9705364; hep-ph/9911383; A. A. Andrianov, D. Espriu and R. Tarrach, *Nucl. Phys.* **B533**, 429 (1998) .
8. M. Golterman and S. Peris, *JHEP* **01**, 028 (2001) .
9. S. R. Beane, *Phys. Rev.* **D64**, 116010 (2001).
10. Yu. A. Simonov, *Phys. Atom. Nucl.* **65**, 135 (2002).
11. J. Gasser and H. Leutwyler, *Nucl. Phys.* **B250**, 465 (1985).
12. M. Golterman and S. Peris, *JHEP* **0101**, 028 (2001); O. Catà, M. Golterman and S. Peris, hep-ph/0506004.
13. Particle Date Group (S. Eidelman *et al.*), *Phys. Lett.* **B592**, 1 (2004).

Future Perspectives

GENERALIZED PARTON DISTRIBUTIONS

M. GUIDAL

Institut de Physique Nucléaire d'Orsay
91405, FRANCE
E-mail: guidal@ipno.in2p3.fr

In the first section, the subject of Generalized Parton Distributions (GPDs) is briefly reviewed. Then, in the second section, we discuss duality considerations which can be used to constrain the parametrisation of the GPDs.

1. Generalized Parton Distributions

In the last decade, Mueller et al. [1], Ji [2] and Radyushkin [3] have shown that the leading order perturbative QCD amplitude for Deeply Virtual Compton Scattering (DVCS) in the forward direction can be factorized, in the Bjorken regime (i.e., in simplifying, large Q^2, where $-Q^2$ is the squared mass of the virtual photon) in a hard scattering part, exactly calculable in pQCD or QED, and a nonperturbative nucleon structure part. This is illustrated in Fig. 1a). In these so-called "handbag" diagram, the lower blob represents the unknown structure of the nucleon and can be parametrized, at leading order pQCD, in terms of 4 generalized structure functions, the Generalized Parton Distributions (GPDs). Using Ji's notation, these are called $H, \tilde{H}, E, \tilde{E}$, and depend upon three variables : x, ξ and t. One considers the process in a frame where the proton has a large momentum along a certain direction which defines the longitudinal components.

$x + \xi$ is the longitudinal momentum fraction carried by the initial quark struck by the incoming virtual photon. Similarly, $x - \xi$ relates to the final quark going back in the nucleon after radiating the outgoing photon. The difference in the longitudinal momentum fraction between the initial and final quarks is therefore -2ξ. In comparison to -2ξ which refers to purely *longitudinal* degrees of freedom, t, the squared 4-momentum transfer between the final nucleon and the initial one, contains *transverse* degrees of freedom (so-called "$k_\perp$") as well.

Intuitively, the GPDs represent the probability amplitude of finding a quark in the nucleon with a longitudinal momentum fraction $x - \xi$ and of putting it back into the nucleon with a longitudinal momentum fraction $x + \xi$ plus some transverse momentum "kick", which is represented by t.

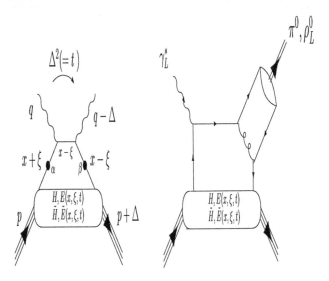

Figure 1. "Handbag" diagrams : a) for DVCS (left) and b) for meson production (right).

The GPDs actually reveal a "double" nature : since negative momentum fractions are identified with with antiquarks, one can define two regions according to whether $|x| > \xi$ or $|x| < \xi$. In the region $-\xi < x < \xi$, one "leg" in Fig. 1a), represents a quark, and the other an antiquark. In this region, the GPDs behave like a meson distribution amplitude and can be interpreted as the probability amplitude of finding a quark-antiquark pair in the nucleon. This kind of information on $q\bar{q}$ configurations in the nucleon and, more generally, the correlations between quarks (or antiquarks) of different momenta, all, being information carried in the concept of GPDs, are completely unknown at the time being, and reveal the richness and novelty of the GPDs.

H and E are independent of the quark helicity and are therefore called *unpolarized* GPDs, whereas $\tilde{H}$ and $\tilde{E}$ are helicity dependent and are called *polarized* GPDs. The GPD's $H^q, E^q, \tilde{H}^q, \tilde{E}^q$ are defined for a single quark flavor ($q = u, d$ and s). The GPDs H and $\tilde{H}$ are actually a generalization of the parton distributions measured in Deep Inelastic Scattering (DIS). Indeed, in the forward direction, H reduces to the quark distribution and

$\tilde{H}$ to the quark helicity distribution measured in DIS.

Furthermore, at finite momentum transfer, there are model independent sum rules which relate the first moments of these GPDs to the elastic form factors. For one quark flavor q, one has :

$$\int_{-1}^{+1} dx H^q(x, \xi, t) = F_1^q(t) , \qquad \int_{-1}^{+1} dx E^q(x, \xi, t) = F_2^q(t) , \qquad (1)$$

$$\int_{-1}^{+1} dx \tilde{H}^q(x, \xi, t) = g_A^q(t) , \qquad \int_{-1}^{+1} dx \tilde{E}^q(x, \xi, t) = h_A^q(t) . \qquad (2)$$

where F_1 and F_2 are related to the nucleon electromagnetic form factors and g_A and h_A denote the axial and pseudoscalar form factors of the nucleon.

It has been shown [4,5,6] that the t dependence of the GPDs can be related, via a Fourier transform, to the transverse spatial distribution of the partons in the nucleon. At $\xi=0$, the GPD$(x, 0, t)$ can be interpreted as the probability amplitude of finding in a nucleon a parton with *longitudinal* momentum fraction x at a given *transverse* impact parameter, related to t. In this way, the information contained in a traditionnal parton distribution, as measured in inclusive DIS, and that contained within a form factor, as measured in elastic lepton-nucleon scattering, are now combined and correlated in the GPD description [7].

The second moment of the GPDs is relevant to the nucleon spin structure [2] :

$$J_q = \frac{1}{2} \int_{-1}^{+1} dx\, x\, [H^q(x, \xi, t = 0) + E^q(x, \xi, t = 0)] , \qquad (3)$$

where J_q is the total (intrinsic spin + orbital momentum) quark contribution to the nucleon total spin. Since the intrinsic spin contribution has been measured through polarized DIS experiments (it is about 20%) and the gluon contribution (J_g) is currently being measured at COMPASS and RHIC, a measurement of the sum rule of Eq.(3) in terms of the GPD's provides a model independent way of determining the quark orbital contribution to the nucleon spin, and therefore complete the "spin-puzzle" : $\frac{1}{2} = J_q + J_g$.

The GPDs reflect the structure of the nucleon independently of the reaction which probes the nucleon. They can also be accessed through the hard exclusive electroproduction of mesons -π^0, ρ^0, ω, ϕ, etc.- (see Fig. 1b)) for which a QCD factorization proof was given in Ref. [8]. In this case, the perturbative part of the diagram involves a 1-gluon exchange and therefore the strong running coupling constant, whose behavior at low energy scales is not fully controlled, makes the calculations and the interpretation of the data more complicated.

We refer to Refs. [9,10] for very complete recent reviews of the field and more details on all these aspects which cannot be covered in this short contribution.

2. Duality aspects

In this section, we discuss duality aspects, the subject of this workshop. We summarize below the ideas developed in Ref. [11]. The correlation between the power behavior of form factors and the behavior of inclusive structure functions $W(x_B)$ of deeply inelastic scattering at large Bjorken variable x_B is a rather popular subject. There are several "duality" ideas. The Bloom-Gilman (BG) idea [12] is that the W^2-integral (where W is the invariant mass of the γ^*-p system) of the hadron contribution is equal to the x-integral of the structure function $W_1(x)$ over a certain W duality region. It originates from the fact that, for sufficiently large x_B, one approaches the exclusive single hadron pole whose contribution is given by the form factor squared multiplied by $\delta(W^2 - m_h^2)$. This gives a relation between the power ν specifying the $(1 - x)^\nu$ behavior of the structure function $W_1(x)$ in the $x \to 1$ region and the power-law behavior of the squared elastic form factor: $F^2(t) \sim (1/|t|)^{\nu+1}$. In the proton case, with usually adopted value $\nu = 3$, one obtains a dipole behavior for the Dirac $F_1(t)$ form factor.

The Drell-Yan[13] (DY) relation expresses an(other) x integral of the structure function in terms of the *first power* of the form factor : if the parton density behaves like $(1 - x)^\nu$, then the relevant form factor should decrease as $1/t^{(\nu+1)/2}$ for large t. The origin of the DY formula lies in the analysis of the large t limit of the light-cone formalism where the form factor can be written as a convolution of light-cone wave functions.

Accidentally, both relations give the same correlation between the two powers (this why they are confused sometimes). In both BG and DY relations, the powers themselves are not fixed ; what is fixed is the general form of the relation between the form factor and the structure function. Perturbative QCD allows to predict definite powers for the asymptotic behavior of form factors and the $x \to 1$ behavior of parton distributions. For example, it gives $(\alpha_s/|t|)^{n-1}$ for a spin-averaged form factor of an n-quark hadron, and it also predicts fixed powers $\alpha_s^{2n-2}(1 - x)^{2n-3}$ for the $x \to 1$ behavior of its valence quark distributions (see Ref. [14]). In the following, we summarize the features of the GPD model published in Ref. [11], which explicitly uses the (DY) duality idea and the relation it imposes between the $x \to 1$ behavior of the structure functions and the t-dependence of elastic form factors, to powerfully constrain the GPD parametrization.

The model implies the dominating role of the Feynman-Drell-Yan mechanism for the hadronic form factors and it is shown in Ref. [11] that a simple way to take into account a Regge behavior needed at small x and, in the same time, satisfy the DY relation[13] is to parametrize the GPD H as :

$$H^q(x,t) = q_v(x)x^{-\alpha'(1-x)t} . \qquad (4)$$

For E, one notes that, experimentally, the proton helicity flip form factor $F_2(t)$ has a faster power fall-off at large t than $F_1(t)$. This means that the $x \sim 1$ behavior of the functions $E(x)$ and $H(x)$ should be different. To produce a faster decrease with t, the $x \sim 1$ limit of the density $E^q(x)$ should have extra powers of $1 - x$ compared to that of $H^q(x)$. In order not to introduce too many free parameters, an idea is to simply multiply the valence quark distributions in the ansatz for $E^q(x)$ by an additional factor $(1 - x)^{\eta_q}$, i.e.,

$$E^u(x) = \frac{\kappa_u}{N_u}(1 - x)^{\eta_u} u_v(x) \qquad \text{and} \qquad E^d(x) = \frac{\kappa_d}{N_d}(1 - x)^{\eta_d} d_v(x) , \quad (5)$$

where N_u and N_d are parameter independent normalization factors ($\kappa_q \equiv \int_0^1 dx\, E^q(x)$). The powers η_u and η_d are to be determined from a fit to the nucleon form factor data and are, with α' of formula 4 the only free parameters of this model.

In Figs. 2, 3, we show the proton and neutron Sachs electric and magnetic form factors. One observes that the model gives a rather good description of all available form factor data for both proton and neutron in the whole t range using the parameter for the Regge trajectory $\alpha' = 1.105$ GeV^{-2}, and the following values for the coefficients governing the $x \to 1$ behavior of the E-type GPDs : $\eta_u = 1.713$ and $\eta_d = 0.566$.

This Regge model reproduces the DY powers for the form factors at large $-t$, and is able to accurately describe existing data. The two parameters η_u and η_d, in particular, allow to describe the decreasing ratio of G^p_E/G^p_M with increasing momentum transfer, as follows from the recent JLab polarization experiments [15,16,17]. Our parametrization leads to a zero for G^p_E at a momentum transfer of $-t \simeq 8$ GeV^2, which will be within the range covered by an upcoming JLab experiment [18].

We end this article by showing an application of this model ; since the GPD E enters the sum rule for the total angular momentum J^q carried by a quark of flavor q in the proton (Eq. 3), our parametrization, in which the

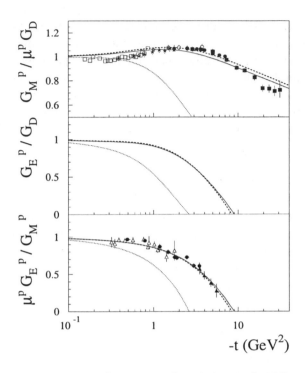

Figure 2. Proton magnetic (upper panel) and electric (middle panel) form factors relative to the dipole form $G_D(t) = 1/(1 - t/0.71)^2$, as well as the ratio of both form factors (lower panel). The solid curves correspond to the Regge model described in the text, with $\alpha' = 1.105$ GeV^{-2}, $\eta_u = 1.713$ and $\eta_d = 0.566$. See Ref [11] for the description of the other curves and the references of the experimental data.

$x \to 1$ limit of E is determined from the F_2^p/F_1^p form factor ratio, allows to evaluate it.

	M_2^q (MRST2002)	$2J^q$ (Regge model)	$2J^q$ (lattice [20])
u	0.40	0.63	0.734 ± 0.135
d	0.22	-0.06	-0.085 ± 0.088
s	0.03	0.03	
$u+d+s$	0.65	0.60	0.65 ± 0.16

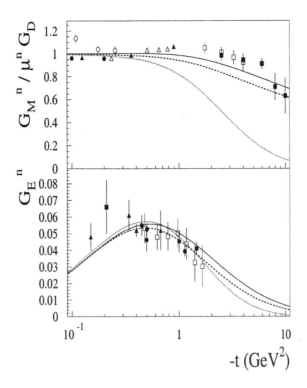

Figure 3. Neutron magnetic form factor relative to the dipole form (upper panel), and neutron electric form factor (lower panel), with curve conventions as in Fig. 2.

In Table 1 (where $M_2^q \equiv \int_{-1}^{1} dx\, x\, H^q(x,0,0)$, i.e. is the total fraction of the proton momentum carried by a quark of flavor q and is therefore already known from the forward parton distributions), we show the values of the quark momentum sum rule M_2^q at the scale $\mu^2 = 1$ GeV^2, using the MRST2002 parametrization [19] for the forward parton distributions. We also show the estimate for J^u, J^d, and J^s at the same scale. Our estimates lead to a large fraction (63 %) of the total angular momentum of the proton carried by the u-quarks and a relatively small contribution carried by the d-quarks. As the d-quark intrinsic spin contribution is known to be relatively large and negative ($\Delta d_v \simeq -0.25$), the small total angular momentum contribution J^d of the d-quarks which follows from our parametrization implies an interesting cancellation between the intrinsic spin contribution and the orbital contribution L^d (with $2J^q = \Delta q + 2L^q$), which should therefore be of size $2L^d \simeq 0.2$. For the u-quark on the other hand, the parametrization yields only a small value for $2L^u$, as our estimate for $2J^u$ is quite close to the intrinsic spin contribution $\Delta u_v \simeq 0.6$. Such a picture is also supported by a recent quenched lattice QCD calculation [20]. for the valence quark contributions to $2J^u$ and $2J^d$. One indeed sees from Table 1 (third column) that the quenched lattice QCD calculation yields quite similar values for $2J^u$ and $2J^d$ as our parametrization R2. It remains to be seen however how large is the sea quark contribution to the GPD E which can enter the spin sum rule of Eq. (3). This sea quark contribution is absent in the quenched lattice QCD calculations of Ref. [20]. The sea quark contribution is also not constrained by the form factor sum rules considered in this paper, which only constrain the valence quark distributions. Ongoing measurements of hard exclusive processes, such as deeply virtual Compton scattering, provide a means to address this question in the near future.

References

1. D. Muller, D. Robaschik, B. Geyer, F. M. Dittes and J. Horejsi, *Fortsch. Phys.* **42**, 101 (1994).
2. X. Ji, *Phys.Rev.Lett.* **78**, 610 (1997); Phys.Rev. **D55**, 7114 (1997).
3. A.V. Radyushkin, *Phys.Lett.* **B380**, 417 (1996); *Phys.Rev.* **D56**, 5524 (1997).
4. M. Burkardt, *Phys. Rev.* **D62**, 071503 (2000) [Erratum-ibid. **D66**, 119903 (2002)] ; *Int. J. Mod. Phys.* **A18**, 173 (2003).
5. M. Diehl, *Eur. Phys. J.* **C25**, 223 (2002) [Erratum-ibid.**C31**, 277 (2003)].
6. J. P. Ralston and B. Pire, *Phys. Rev.* **D66**, 111501 (2002).
7. A. V. Belitsky, X. d. Ji and F. Yuan, *Phys. Rev.* **D69**, 074014 (2004).
8. J.C. Collins, L. Frankfurt and M. Strikman, *Phys. Rev.* **D56**, 2982 (1997).
9. K. Goeke, M. V. Polyakov and M. Vanderhaeghen, *Prog. Part. Nucl. Phys.* **47**, 401 (2001).
10. M. Diehl, *Phys. Rept.* **388**, 41 (2003).
11. M. Guidal, M. V. Polyakov, A. V. Radyushkin and M. Vanderhaeghen, *Phys. Rev.* **D72**, 054013 (2005).
12. E. D. Bloom and F. J. Gilman, *Phys. Rev. Lett.* **25**, 1140 (1970).
13. S. D. Drell and T. M. Yan, *Phys. Rev. Lett.* **24**, 181 (1970).
14. G. P. Lepage and S. J. Brodsky, Phys. Rev. D **22**, 2157 (1980).
15. M. K. Jones et al. [Jefferson Lab Hall A Collaboration], *Phys. Rev. Lett.* **84**, 1398 (2000) [arXiv:nucl-ex/9910005].
16. O. Gayou et al., *Phys. Rev.* **C64**, 038202 (2001).
17. O. Gayou et al. [Jefferson Lab Hall A Collaboration], *Phys. Rev. Lett.* **88**, 092301 (2002).
18. JLab experiment E-01-109 / E-04-108, spokespersons : E. Brash, M. Jones, C. Perdrisat, V. Punjabi.
19. A. D. Martin, R. G. Roberts, W. J. Stirling and R. S. Thorne, *Phys. Lett.* **B531**, 216 (2002).
20. M. Gockeler, R. Horsley, D. Pleiter, P. E. L. Rakow, A. Schafer, G. Schierholz and W. Schroers [QCDSF Collaboration], *Phys. Rev. Lett.* **92**, 042002 (2004).

EXPERIMENTAL OVERVIEW OF EXCLUSIVE PROCESSES

D. HASCH

INFN-Laboratori Nazionali di Frascati,
via E. Fermi, 40,
Frascati (RM), 00044, Italy
E-mail: delia.hasch@lnf.infn.it

Hard exclusive leptoproduction of real photons and mesons provides access to the unknown Generalised Parton Distributions (GPD) of the nucleon which give a unified description of hadronic structure. Different observables, like cross sections, single-spin asymmetries or angular distributions for deeply virtual exclusive real photon and meson production have been measured at HERA (HERMES, H1 and ZEUS), CERN SPS (COMPASS) and JLab (CLAS). The recent experimental results will be reviewed and compared to the GPD expectations. Finally, prospects of future measurements will be presented.

1. Introduction

Interest in the hard exclusive electroproduction of mesons and photons has grown since a QCD factorisation theorem was proven [1]. The Generalised Parton Distributions (GPDs) appearing in the factorisation scheme to parametrize the target nucleon provide a unified description of exclusive and inclusive reactions. The quantum numbers of the produced particle select different GPDs. While exclusive vector meson production is sensitive to unpolarised GPDs (H and E), pseudoscalar meson production is sensitive to polarised ones ($\tilde{H}$ and $\tilde{E}$) without the need for a polarised target or beam. Deeply Virtual Compton Scattering (DVCS) is sensitive to all four distributions. With the measurements of different observables, like cross sections and single-spin asymmetries, one can select different combinations of GPDs.

At high virtuality Q^2 of the exchanged photon, the measurments of exclusive processes are very challenging. Precision measurements require high luminosity$\times$acceptance since the cross sections are small and high resolutions and/or complete reconstruction of the reaction products to ensure the exclusivity of the process.

In this paper, the latest results obtained at HERA, CERN and JLAB on hard exclusive processes will be presented. At HERA the 920 GeV proton beam colliding with the 27.5 GeV polarised positron or electron beam is used by the H1 and ZEUS experiments while HERMES uses the positron (electron) beam only incident on internal polarised or unpolarised pure gas targets of different types. The COMPASS experiment at CERN SPS uses the polarised 160GeV muon beam scattered on a polarised solid deuterium target (^{6}LiD). At JLAB the 4.2 GeV and 5.75 GeV polarised electron beams incident on a polarised solid proton target (NH$_3$) are are used by the CLAS experiment. None of the experiments can fully reconstract the exclusive events. At H1 and ZEUS the outgoing target proton escapes down the beam pipe but the nearly 4π spectrometers can fully reconstruct the remaining final state and ensure exclusivity. The other experiments (HERMES, CLAS and COMPASS) are limited by the acceptance of their spectrometers. They therefore request the missing mass (missing energy) of the reaction $lN \rightarrow lMX$, where M is a meson or a real photon, to correspond to the nucleon mass (to zero) and the non-exclusive background is subtracted.

2. Deeply Virtual Compton Scattering

The hard exclusive electropoduction of a real photon appears to provide the theoretically cleanest access to GPDs. It is also unique among the hard scattering processes in that DVCS amplitudes (magnitude and phase) can be determined. This is possible through a measurement of the interference between the DVCS and Bethe–Heitler (BH) processes, in which the photon is radiated from a parton in the former and from the lepton in the latter process. These processes have an identical final state, i.e., they are indistinguishable and thus the photon production amplitude τ is given as the coherent sum of the amplitudes of the DVCS (τ_{DVCS}) and BH (τ_{BH}) processes. At leading twist, the BH–DVCS interference term I can be written as

$$I = \pm [c_1^I \cos\phi \, \mathrm{Re}\tilde{M} - P_l \, s_1^I \, \sin\phi \, \mathrm{Im}\tilde{M}], \tag{1}$$

where +(-) denotes a negatively (positively) charged lepton beam with longitudinal polarization P_l. The DVCS amplitude $\tilde{M}$ is given by a linear combination of the nucleon form factors with the so–called Compton form factors [2], which are convolutions of the twist–2 GPDs with the hard scattering amplitude.

In the following, the recent H1 and ZEUS cross section measurements, which access the squared DVCS amplitude, and new results from HERMES and CLAS on azimuthal asymmetry measurements, which access the DVCS amplitude directly via the interference term I, will be presented. Here, The azimuthal angle ϕ is defined as the angle between the lepton scattering plane and the photon production plane.

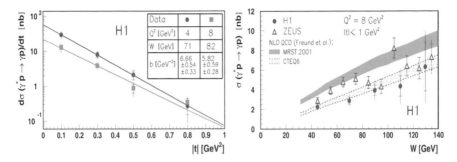

Figure 1. Left: The cross section $\gamma^* p \to \gamma p$ differential in t, for $Q^2 = 4$ GeV^2 and $Q^2 = 8$ GeV^2 measured by H1. Right: The $\gamma^* p \to \gamma p$ cross section as a function of W for $Q^2 = 8$ GeV^2 mesured by H1 and ZEUS compared to NLO QCD predictions based on MRST 2001 and CTEQ6 PDFs. The band associated with each prediction corresponds to the uncertainty on the measured t-slope.

The differential cross section in t, where t is the square of the four-momentum transfer to the target, is measured by H1 at two different Q^2 values as shown in Fig. 1 left hand side. The t dependence is parametrised as $e^{-b|t|}$. Combining the two data sets, the t slope is measured to be $b = 6.02 \pm 0.35(\text{stat}) \pm 0.39(\text{sys})$ GeV^{-2} for $Q^2 = 8$ GeV^2 and $W = 82$ GeV. The right hand side of Fig. 1 presents the cross section measured by H1 and ZEUS as a function of W for $Q^2 = 8$ GeV^2. The steep rise of the cross section with W indicates the presence of a hard scattering process. Fig. 1 also compares the measurements with QCD predictions calculated at NLO by Freund and McDermott [3]. The theoretical estimates agree well with the data for both shape and absolute normalisation. The uncertainty in the normalization for the theory is significantly reduced owing to the H1 measurement of the cross section (exponential) t slope; this uncertainty becomes smaller than the input PDF uncertainty which is quantified comparing MRST and CTEQ PDF set based predictions.

In contrast to the squared BH and DVCS amplitudes, the interference term I (see Eq. 1) does depend on the sign of the beam charge. Therefore a measurement of a cross section asymmetry with respect to the beam charge can isolate the real part of the interference term, while the imaginary part can be isolated with a polarized lepton beam ($P_l \neq 0$). Measurements of

224

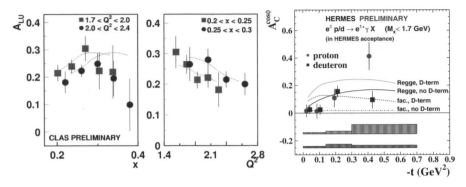

Figure 2. *Left panel: Beam-spin asymmetry A_C for hard electroproduction of photons as a function of x and Q^2 measured by CLAS. Right panel: Beam-charge asymmetry A_C for hard electroproduction of photons as a function of $-t$ measured by HERMES. The curves correspond to different GPD parametrisations (see text).*

azimuthal asymmetries with respect to the beam spin, accessing the imaginary part of $\tilde{M}$ via a $\sin\phi$ modulation, have been reported on hydrogen [4,5] as well as on deuterium and neon [6]. New preliminary data on the beam-spin asymmetry from CLAS with improved statistics allow to study the kinematic dependences in more detail as shown in Fig. 2 left hand side.

The first extraction of an azimuthal asymmetry with respect to the beam charge, accessing the real part of $\tilde{M}$ via a $\cos\phi$ modulation, has been performed by Hermes using hydrogen and deuterium targets and is presented in the right hand side of Fig. 2 as function of $-t$. Also shown are GPD model based calculations [7,8] carried out at the average kinematics of each $-t$ bin. It is apparent that the beam-charge asymmetry is particularly sensitive to different GPD models. The data, representing a tiny sample of about 10pb^{-1} taken with the electron beam and hydrogen and deuterium targets in 1998, do not yet permit to distinguish between the models. However, since 2005 HERMES is taking again data with an electron beam in order to increase the statistical accuracy of the beam-charge asymmetry measurement. Both the beam-charge and beam-spin asymmetries are especially sensitive to the GPD H as contributions from other GPDs are kinematically suppressed in the energy range of current experiments (HERMES and CLAS).

The single-spin asymmetry w.r.t. to a longitudinally polarised target has been reported by CLAS and HERMES. For this observable the GPDs H and $\tilde{H}$ enter with a similar kinematic weight permiting to constrain mod-

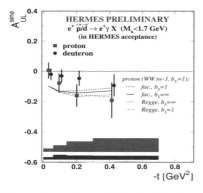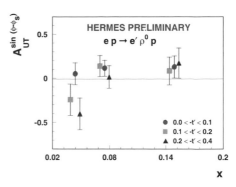

Figure 3.   *Left panel: longitudinal target-spin asymmetry A_{UL} for hard electroproduction of photons as a function of $-t$. The curves correspond to different GPD parametrisations (see text). Right panel: transverse target-spin asymmetry A_{UT} for hard electroproduction of ρ^0 as a function of x for three different $-t$ ranges.*

els for the imaginary part of the GPD $\tilde{H}$. The longitudinal target-spin asymmetry obtained by HERMES using hydrogen and deuterium targets is shown in the left hand side of Fig. 3 as function of $-t$ together with GPD model calculations [8].

By the time of this proceedings, an extremly exiting observable has been measured by HERMES with a transversely polarised hydrogen target. This single target-spin asymmetry is sensitive to the GPD E and may therefore allow to constrain the total angular momentum of u-quarks J^u for the first time [9,10].

3. Exclusive Meson Production

Exclusive meson production provides valuable information about the flavour dependence of GPDs.

For meson electroproduction the factorisation of the process has been demonstrated for longitudinal virtual photons only. However, the transverse contribution to the cross section is predicted to be suppressed by a power of $1/Q^2$ with respect to the longitudinal one which is expected to fall like $1/Q^6$. For pseudoscalar meson production the separation of the longitudinal and transverse parts of the cross sections is not feasibile in current experiments. Therefore the test of the Q^2-dependence of the measured cross sections being in agreement with the prediction is essential.

In contrast, for vector meson production the separation of longitudinal and transverse cross sections is possible using the angular dependences of the

226

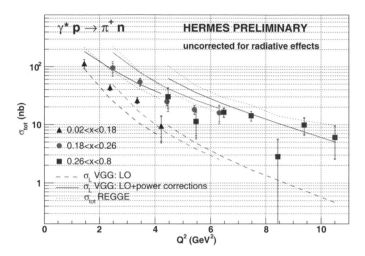

Figure 4. *Total cross section for π^+ production as a function of Q^2 for three different x ranges and integrated over t. The curves represent calculations based on a GPD and Regge model.*

decaying products.

Results from HERMES on the W-dependence of the ρ_L^0 cross section for different Q^2 and on the cross section ratio ρ_L^0/ϕ_L have been reported earlier and compared to GPD model calculations. Also CLAS reported results for ρ_L^0 and ϕ_L cross section measurements however at low $W < 2$ GeV obtained with the 4.2 GeV beam. Data taken with higher beam energy (5.75 GeV) are currently analysed.

First results from COMPASS on the extraction of few ρ^0 spin density matrix elements in an one dimensional projection of the angular distributions are available and the ratio $R = \sigma_L/\sigma_T$ has been calculated assuming S-channel helicity conservation. This analysis will be extended towards the ρ_L^0 cross section extraction.

An exiting observable in ρ^0 production has been measured by HERMES using data with transverse target polarisation. Here, the interference between the scalar (E) and vector (H) amplitudes leads to a single-spin asymmetry which is sensitive to the total angular momentum contribution $J^{u,d}$ for the u, d-quarks to the proton spin. Fig. 3 right hand side shows the first measurment of the transverse target-spin asymmetry for ρ^0 as function of x for different $-t$ ranges. The data exhibit a large negative asymmetry at low x and large $-t$ and the same x-dependence behaviour as predicted by GPD models [8].

HERMES has measured the total cross section for exclusive π^+ production shown in Fig. 4 as function of Q^2 for three different x-ranges. The data is compared to calculations for the longitudinal part of the cross section computed from a GPD model [7] The Q^2 dependence is in general agreement with the theoretical expectation. While the leading-order calculations (full lines) underestimate the data, the evaluation of the power corrections (dashed lines) appears too large. The data have also been compared to calculations for the total cross section computed by a Regge model [11] (dotted lines). These calculations overestimate the data. Within this model, the longitudinal and transverse part of the cross section have been calculated demonstrating the transverse part is suppressed at least by a factor of four compared to the longitudinal part.

4. Conclusions and Future Measurements

Generalised Parton Distributions have merged as a powerful, attractive and unifying concept for the nucleon structure. Data on the hard exclusive production of real photons and mesons are already available from experiments at HERA, JLAB and CERN. These results provide very first constraints to GPD models.

New extensive data sets on exclusive meson production will become available soon from HERMES, COMPASS and CLAS. Dedicated DVCS experiments at JLAB (HallA and CLAS) will measure beam-spin asymmetries for hydrogen, detecting all three particles in the final state to ensure full exclusivity.

At the end of 2005, HERMES will install a recoil detector around the gas target in order to detect the recoil proton in the DVCS process. This will allow to fully reconstruct the final state and to distinguish DVCS events from associated DVCS with a resonance in the final state. It is planned to measure the $-t$ dependence of the beam-spin and beam-charge asymmetries for hydrogen and heavier nuclei with high precision. Data taken with the recoil will also allow to measure exclusive production of neutral pseudoscalar mesons on hydrogen. In particular, the study of the ratio for different pseudoscalar mesons (π^+ over π^0, π^0 over η, ...) will provide information on the different contributions with respect to the nature of the produced mesons.

A global analysis of all observables in hard exclusive meson and real photon production is necessary in order to contrain the models of GPDs and to extract detailed information about the nucleon structure.

228

References

1. J.C. Collins, L. Frankfurt, M. Strikman, Phys. Rev. **D56**(1997),2982.
2. A.V. Belitsky, D. Müller and A. Kirchner, Nucl. Phys. **B629** (2002) 323.
3. A. Freund, M. McDermott and M. Strikman, Phys. Rev. **D67** (2003) 036001.
4. HERMES Coll., A. Airapetian et al., Phys. Rev. Lett. **87** (2001) 182001.
5. CLAS Coll., S. Stepanyan et al., Phys. Rev. Lett. **87** (2001) 182002.
6. F. Ellinghaus, R. Shanidze and J. Volmer (for the HERMES Coll.), hep-ex/0212019.
7. M. Vanderhaeghen, P.A.M. Guichon and M. Guidal, Phys. Rev. **D60** (1999) 094017.
8. K. Goeke, M.V. Polyakov and M. Vanderhaeghen, Prog. Part. Nucl. Phys. **47** (2001) 401.
9. HERMES Collaboration (Z. Ye et al.), to appear in the proceedings of The European Physical Society International Conference on High Energy Physics, Lisboa, Portugal, July 21th - 27st.
10. F. Ellinghaus, W.-D. Nowak, A.V. Vinnikov, Z. Ye, hep-ph/0506264.
11. J.M. Laget, Phys. Rev. **D70** (2004),054023; *Private communication.*

TRANSVERSE POLARIZATION AND QUARK-GLUON DUALITY *

O. V. TERYAEV

Joint Institute for Nuclear Research, Dubna, 141980 Russia

The possible advantages of transverse polarized DIS for applications of QCD duality are analyzed. The status of Bloom-Gilman duality in QCD is discussed and its validity for the structure function g_T is deduced. The special role of this function is also manifested in the analysis of GGDH sum rules.

1. Introduction. While the leading spin structure of nucleon in hard processes is described by structure function $g_1(x)$ corresponding to its kinematically enhanced longitudinal polarization, both chiral even $g_T^a(x)$, and chiral-odd (transversity) $h_1^a(x)$ distributions are required to describe the transverse polarization. Although the later in the case of DIS formally corresponds to subleading twist 3, its contribution, like that of the various higher twist (HT) distributions are very important when the transition region (with the typical scales of order of 1 GeV) is considered.

Despite the kinematical suppression and the dynamical complexity of transverse polarization in the helicity basis it has some advantage. Namely, the natural definition of invariant amplitudes involves the tensors with the coefficients $g_T = g_1 + g_2$ and g_2,

$$W_A^{\mu\nu} = \frac{-i\epsilon^{\mu\nu\alpha\beta}}{pq} q_\beta (g_1(x, Q^2) s_\alpha + g_2(x, Q^2)(s_\alpha - p_\alpha \frac{sq}{pq})) =$$
$$\frac{-i\epsilon^{\mu\nu\alpha\beta}}{pq} q_\beta ((g_1(x, Q^2) + g_2(x, Q^2)) s_\alpha - g_2(x, Q^2) p_\alpha \frac{sq}{pq}), \qquad (1)$$

so that only g_T contributes in the case of transverse polarization while both contribute in the case of longitudinal one, It is just the adopted definitions of invariant formfactors leading to the (approximate) cancellation of g_2 and providing the contact with helicity formalism. As soon as just the invariant

---

*This work is partially supported by grant RFBR 03-02-16816.

amplitudes are the natural objects for the application of QCD duality, the most developed technique being represented by QCD Sum Rules (SR) [1]. Historically, such a special role of transverse polarization was first observed in the analysis if Generalized Gerasimov-Drell-Hearn(GGDH)SR.

2. Generalized GDHSR: role of g_T

The GGDH SR represents the situation when all the higher twists contributions are important. While at large Q^2 the formfactor

$$I_1(Q^2) = \frac{2M^2}{Q^2}\Gamma_1(Q^2) \equiv \frac{2M^2}{Q^2}\int_0^1 g_1(x, Q^2)dx \qquad (2)$$

behaves like $1/Q^2$, for lower Q^2 the higher powers of $1/Q^2$ are manifested. At the same time, their infinite sum is required to get the GDH value at $Q^2 = 0$ (provided the elastic contribution is subtracted): $I_1(0) = -\mu_A^2/4$. As the large Q^2 value for proton is positive, the corresponding I_1 must represent the dramatic structure. It was suggested[2] that it emerges from the elastic contribution to similar formfactor I_2, related to g_2 by the way analogous to (2). This elastic contribution is in fact controlled by the Burkhardt-Cottingham sum rule which was recently checked experimentally at low Q^2 [3]. At the same time, the special role of transverse polarization was suggested. Namely, similar formfactor I_T was assumed to be smooth, which is natural from the point of view of QCD SR. Indeed, its real photon limit $I_T(0) = -\mu_A/4$ is linear in μ_A and one may expect to get the corresponding low energy theorem from the Ward identities, analogously to, say, low energy theorem for pion formfactor (see[2,4] and Ref. therein). The model for I_T was first chosen[2] in the simplistic way neglecting all the possible corrections, and this allowed to predict the low crossing point for I_1. At the same time, the current high accuracy of JLAB data required to take into account the radiative and power corrections[5]. This leads to following behaviour (Fig. 1a).

Implementing the smooth model[6] for I_T^{p-n} which is also linear function of μ_A^p, one easily gets the similar expression[5] for neutron (Fig. 1b). The results are in good agreement with the recent JLAB data and the results of the resonance approach of Burkert and Ioffe [3]. The latter agreement may be the another manifestation of the fact that g_T is a good candidate for BG duality.

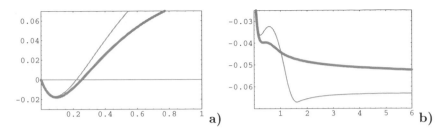

Figure 1. a) The proton Γ_1 as a function of Q^2. b) The neutron Γ_1 as a function of Q^2. Solid lines correspond to perturbative and power corrections included.

3. Bloom-Gilman duality: for which structure function it holds better?

Let us elaborate the role of transverse polarization in the analysis of the Bloom-Gilman (BG) duality [7]. The advantage of structure function g_T [8] may be also seen from the following consideration. Starting with invariant structures g_T and g_2 one should recall the strong Q^2 dependence of BCSR which is one of the basic ingredients of the approach to GGDHSR in the previous section. The same property is preventing the BG duality to be valid for g_2. Indeed, the basic feature of BG duality is "sliding" [7] of resonances under the smooth DIS curve when Q^2 is changed. However, rapid variation of the integral over the resonances, absent for the smooth curve, prevents from such a behaviour. One should therefore expect that the other invariant tensor, corresponding to the structure function g_T, should be better candidate for BG duality. This reasoning, when applied to the spin-averaged structure function, should lead to the better validity of BG duality for F_2, whose integral is protected by momentum sum rule. This suggestion is actually supported by the data, as it is known that BG duality for g_1 structure function is violated by the contribution of $\Delta(1232)$ resonance [9]. At the same time, its dominant magnetic transition formfactor contributes only to g_T [4]. The recently found [3] "mirror" behaviour of g_1 and g_2 at low Q^2 should be also considered as the manifestation of the special role of g_T. The analysis of g_T using the modern advanced methods [10] is highly desirable.

Let us discuss the quantitative description of BG duality which requires the account [11] for the large x enhanced HT terms behaving like $(M^2/(1-x)Q^2)^n \approx (M^2/s)^n$. The quantitative analysis of these terms may be performed by exploring the technique of Borel sum rules (SR), which is very popular than vacuum power corrections are considered [1], but, to my knowledge, never applied before for studies of BG duality. As soon as only these enhanced power corrections are considered, the Borel SR in the variable s is especially convenient. Comparing the analysis of BG duality to the most simple case of static meson characteristics one may see the two complications. Namely, the BG problem contains two scales, s and Q^2, and the sum rule contains the contributions from s and u channels. However, keeping of the leading (in $1/(1-x)$) power corrections allows to avoid both complications. Calculating the Compton amplitude in the asymmetric point in the non-physical region close to the s threshold, one may keep only the enhanced power corrections and neglect the contribution of u channel. As a result, the Borel sum rule is completely similar to the case of meson characteristics. Assuming the ansatz for spectral density

$$\rho(s) = \theta(s - s_0)\rho^{pert}(s) + \theta(s_0 - s)\rho^{Res}(s), \qquad (3)$$

where s_0 is the duality interval, and putting the Borel parameter to infinity, which leads to the disappearance of the power corrections, one gets

$$\int_{s_{min}}^{s_0} ds(\rho^{pert}(s) - \rho^{Res}(s)) = 0, \qquad (4)$$

which is just BG duality. Note that the calculation of s_0 from QCD, which is the real problem of QCD SR , would require the explicit account for the enhanced HT terms. This problem was already studied in detail [12] in the case of spin-averaged structure function F_2. One may worry, whether this analysis, assuming very large x, is applicable for studies of BG duality, corresponding to lower x. The positive answer may result from the following simple reasoning. The large x behavior is governed by the power dependence $(1-x)^b$ and the exponent b, once established at very large x, should govern also the behaviour at lower ones, relevant for BG duality. The similar reasoning may explain, why BFKL asymptotics, requiring very small x, may be successfully applied (see [13] and Ref. therein) to describe data at much larger x. Namely, the Regge behaviour x^a, once established at very low x, should also be applicable, with reasonable numerical accuracy, to much larger ones. It would be also interesting to include such "enhanced" HT contributions to the current analysis of experimental data, where their trace may have been already seen [14].

4. Duality in SIDIS

. Let us briefly discuss two relatively new aspects of QCD duality and similarity to QCD sum rules approach in Semi-Inclusive DIS (SIDIS). Here the higher twist mass parameter contributions may be compensated by various kinematical variables, and terms $(M/p_T)^n$ are of most importance. It is these terms which are taken into account by the transverse momentum dependent (TMD) distribution and fragmentation functions. As these terms are not small at $M \sim p_T$ one is often speaking about leading twist effects. However, the analysis in the coordinate space [15] shows that one is rather dealing with the infinite sum of higher twists, completely similar to the one appearing in the case of non-local quark condensates. Their difference with TMD functions is mainly the following: for non-local condensates one is dealing with the vacuum matrix elements, while for TMD functions with the hadron ones. At the same time, the intrinsic transverse momentum plays the same role as the quark virtuality in the case of non-local condensates. Say, the coordinate analog of Collins fragmentation function is of twist-3 nature and corresponds to the weighted k_T moment of the latter. The consideration in the coordinate space provides the the factorization proof for the (weighted) p_T−averaged cross-sections, analogous to the one deduced for Drell-Yan process by Efremov and Radyushkin (see [15] and Ref. therein). The physical essence of such a proof is that the integrated SIDIS cross-section is completely similar to the one of DIS, establishing a sort of SIDIS-DIS duality. This, in fact, may explain the early scaling in SIDIS observed in JLAB [16], which may be considered as a counterpart of "handbag dominance" in DIS.

Another aspect of QCD duality comes when the transition to the low z region is considered. In that case the contribution of correlated distribution and fragmentation functions described by so-called fracture function is important. This notion was originally introduced [17] for diffractive process, while the use of the same "handbag dominance" [15,18] allows to apply it at fixed target experiments. The relevance of correlated distribution and fragmentation in DIS was confirmed by the Monte-Carlo analysis [19]. The crucial point of the analysis of fracture functions is that the momentum sum rule for fragmentation functions

$$\sum_i \int dz z D_i(z) = 1, \tag{5}$$

leads to the completeness of the corresponding description of SIDIS, so that one have the "fragmentation-fracture duality" (c.f. [16]) at low z. Such a duality may explain the success of the description in terms of independent fragmentation, as well as to serve as a guide for modelling the fracture functions.

The notion of fracture function may be naturally extended to include transverse polarization [18,15]. It is interesting that just for transverse asymmetries there is a clear evidence [20] of the breaking of factorization between distribution and fragmentation functions.

234

5. Conclusions. The transverse polarization plays a special role for QCD duality because of its simplicity in terms of invariant, rather than helicity, amplitudes. This is seen in the role played by structure function g_T when the GGDHSR [2,4−6] and BG duality [8] are considered. It happens that the guidelines from QCD SR methods appear to be rather convenient when spin-dependent DIS or SIDIS is considered. Namely, the Borel sum rules provide the complementary description of BG duality. Passing to SIDIS case, the TMD function may be analyzed in coordinate space, where their analogs are quite similar to non-local vacuum condensates. In such an approach the appearance of the resummed infinite set of higher twists, rather than leading twist, is clearly manifested. The duality od SIDIS to DIS provides the reason of early scaling, while the fragmentation-fracture duality is relevant for the description of target fragmentation region.

Finally, I would like to express my deep gratitude to Organizers for warm hospitality and financial support.

References

1. M. A. Shifman, A. I. Vainshtein and V. I. Zakharov, Nucl. Phys. B **147**, 385 (1979).
2. J. Soffer and O. Teryaev, Phys. Rev. Lett. **70**, 3373 (1993).
3. Z.E. Meziani, These Proceedings.
4. J. Soffer and O. Teryaev, Phys. Rev. **D51**, 25 (1995).
5. J. Soffer and O. Teryaev, Phys. Rev. D **70**, 116004 (2004).
6. J. Soffer and O. Teryaev, Phys. Rev. Lett. **B545**, 323 (2002).
7. C. Carlson, These Proceedings,
8. O.V. Teryaev, In Proceedings of 16 International Spin Physics Symposium, Trieste, 2004/
9. S. Simula, M. Osipenko, G. Ricco and M. Taiuti, Phys. Rev. **D65**, 034017 (2002).
10. A. Fantoni, These Proceedings.
11. A. De Rujula, H. Georgi and H. D. Politzer, Phys. Rev. D **15**, 2495 (1977).
12. E. Gardi, G. P. Korchemsky, D. A. Ross and S. Tafat, Nucl. Phys. B **636**, 385 (2002).
13. J. Soffer and O. V. Teryaev, Phys. Rev. D **56**, 1549 (1997).
14. W. Melnitchouk, These Proceedings
15. O. V. Teryaev, arXiv:hep-ph/0310069, Proceedings of DIS-03, p.827; Proceedings of SPIN-03, p. 200.
16. P. Bosted, These Proceedings
17. L. Trentadue and G. Veneziano, Phys. Lett. B **323**, 201 (1994); M. Grazzini, L. Trentadue and G. Veneziano, Nucl. Phys. B **519**, 394 (1998)
18. O. V. Teryaev, Acta Phys. Polon. B **33**, 3749 (2002) [arXiv:hep-ph/0211027].
19. A. Kotzinian, Phys. Lett. B **552**, 172 (2003)
20. A. V. Efremov, Annalen Phys. **13**, 651 (2004)

THE EXTENSION TO THE TRANSVERSE MOMENTUM OF THE STATISTICAL PARTON DISTRIBUTIONS

F. BUCCELLA

Dipartimento di Scienze Fisiche, Università di Napoli,
and INFN, Sezione di Napoli, Italy
Via Cintia, I-80126, Napoli
E-mail: Franco.Buccella@na.infn.it

The extension of the statistical approach to the transverse degrees of freedhom explains a moltiplicative factor, we were obliged to introduce in a previous work to comply with experiment for the Fermi-Dirac functions of the light quarks. It is possible to get light antiquark distributions similar to the ones proposed there.

1. Introduction

I would like to devote this talk to Prof.Bruno Toushek, to whom this anphitheatre is devoted and who taught me statistical mechanics.

In his beautiful talk [1] Jacques has shown the undisputable phenomenological success of the quantum statistical approach proposed by us some years ago [2] with a relationship between the light q and $\bar{q}$ distributions deduced [3] by the chiral properties of QCD.

More recently [4] we stressed the experimental evidence in several structure function for a property typical of the statistical distributions: the change of slope above the highest "thermodynamical potential", X_{0u}^{+}.

The fermion distributions are given by the sum of two terms [2], a quasi Fermi-Dirac function and a helicity independent diffractive contribution equal for all light quarks. The first terms are given by:

$$xq^h(x, Q_0^2) = \frac{AX_{0q}^h x^b}{\exp\left[\frac{x-X_{0q}^h}{\bar{x}}\right] + 1} \tag{1}$$

$$x\bar{q}^h(x, Q_0^2) = \frac{\bar{A}x^{2b}}{X_{0q}^{-h} \exp\left[\frac{x+X_{0q}^{-h}}{\bar{x}}\right] + 1} \tag{2}$$

at the input energy scale $Q_0^2 = 4 GeV^2$.

The universal parameter $\bar{x}$ plays the role of "temperature" and $X_{0q}^{\pm}$ of the "thermodynamical potentials".

The equations contain the spurious $[X_{0q}^h]^\pm$ factors.

Also the quantum statistical approach to the parton distributions is often criticized, since a p_T^2 phase space proportional to Q^2 is expected to imply a dilution giving rise to the Boltzmann limit.

Sometimes in physics it is better to have two problems than just one!

The aim of this talk is to account for these factors by extending the statistical approach to the transverse degrees of freedhom.

To find the most probable occupation numbers n_i for the energy levels ϵ_i of N distinguishible particles with total energy E, one looks with the Lagrange multiplier method for the maximum of:

$$\ln \frac{N!}{\Pi_i n_i!} = +\alpha(N - \sum_i n_i) + \beta(E - \sum_i n_i \epsilon_i) \tag{3}$$

The result is:

$$n_i = \exp(-\alpha - \beta \epsilon_i) \tag{4}$$

with α and β to be determined by:

$$\sum n_i = N \tag{5}$$

$$\sum n_i \epsilon_i = E \tag{6}$$

Statistical considerations have been applied to very high energy p-p scattering [6], by writing for the energy of the produced particles the approximate formula:

$$E_i \simeq p_{zi} + \frac{p_{Ti}^2 + m_i^2}{p_{zi}} \tag{7}$$

which holds when p_{zi} is larger than the transverse mass. By writing $p_{zi} = x_i p_z$, we find the x, p^T dependance:

$$\exp\left(-\frac{x}{\bar{x}} - \frac{p_T^2 + m^2}{x \mu^2}\right) \tag{8}$$

From which we deduce that the ratio $\frac{p_T^2}{x}$ does not depend on the particle considered. Also the property that both those quantities are increasing with the mass, follows from eq.(8). Finally we find [6] for the high p_T behaviour of the produced particles a universal behaviour proportional to $\exp -\frac{p_T}{T_H}$, apart some soft power dependance, again in agreement with experiment.

For the deep inelastic scattering on the proton, one may write:

$$E_P \simeq p_z + \frac{M_P^2}{p_z} \tag{9}$$

$$E_i \simeq p_{zi} + \frac{p_{Ti}^2 + m_i^2}{p_{zi}} \tag{10}$$

In the following, since we will consider the light fermions, we will neglect m_i.

The condition that the the partons carry the momentum and the energy of the proton implies:

$$\sum_i \int_0^1 dx \int_0^{xM^2} xp_i(x, p_T^2) dp_T^2 = 1 \tag{11}$$

$$p_z \sum_i \int_0^1 dx \int_0^{xM^2} xp_i(x, p_T^2) dp_T^2 + \sum_i \int_0^1 dx \int_0^{xM^2} p_i(x, p_T^2) \frac{p_T^2}{2xp_z} dp_T^2$$
$$= p_z + \frac{M^2}{2p_z} \tag{12}$$

where i runs on all the partons (quarks, antiquarks and gluons) and we have assumed that the mass of proton is smaller than its momentum and that the transverse momenta of the partons are smaller than their longitudinal momenta (both the approximations should hold in the deep inelastic regime). The upper limit for the integral on p_T^2, xM^2, has been chosen, as it will be clear in the next formula, to correspond to one parton taking the wholetransverse energy, as the upper limit 1 for the integral on x corresponds to one parton carrying the whole momentum. In the future, to simplify the formulas, we shall take as upper limit ∞, since the contribution of the tail is negligible.

From Eq.(11-12) it is easy to derive the constraint:

$$\sum_i \int_0^{xM^2} p_i(x, p_T^2) \frac{p_T^2}{x} dx = M^2 \tag{13}$$

Eq.'s (11,13) confirm the consistency of the two approximation for the proton and the partons energy: if the partons carry the longitudinal momentum and the energy of the proton and M is smaller than p_z, also the transverse energy of the partons should be in general smaller than the longitudinal one.

From Eq.'s (11,13) one should get in the Boltzmann limit the (x, p_T) distribution:

$$\exp\left(-\frac{x}{\bar{x}} - \frac{p_T^2}{x\mu^2}\right) \tag{14}$$

where $\frac{1}{\bar{x}}$ and $\frac{1}{\mu^2}$ are the Lagrange multipliers associated to the constraints (11) and (13).

The corresponding quantum statistics distributions for the fermions are (we consider the non-diffractive terms and, instead of the helicity, the spin component of the partons along the momentum of the proton):

$$xq^{S_z}(x,p_T) = \frac{f_q(x)}{\exp\frac{x-X_{0q}^{S_z}}{\bar{x}}+1} \frac{1}{\exp\left(\frac{p_T^2}{x\mu^2}-Y_{0q}^{S_z}\right)+1} \tag{15}$$

$$x\bar{q}^{-S_z}(x,p_T) = \frac{f_{\bar{q}}(x)}{\exp\frac{x+X_{0q}^{S_z}}{\bar{x}}+1} \frac{1}{\exp\left(\frac{p_T^2}{x\mu^2}+Y_{0q}^{S_z}\right)+1} \tag{16}$$

It is reasonable to assume for the potentials X and Y the proportionality relationship:

$$Y_{0q}^{S_z} = kX_{0q}^{S_z} \tag{17}$$

It implies that the partons with a larger contribution to their first moments from the non-diffractive part not only have a broader x dependance, but also, at every x, a broader p_T^2 dependance. This assumption has a rather intriguing consequence. In fact when one integrates on p_T^2 the second factor in the right-hand side of Eq.(15), one finds:

$$\int_0^\infty \frac{1}{\exp\left(\frac{p_T^2}{x\mu^2}-Y_{0q}^{S_z}\right)+1} dp_T^2 = x\mu^2 R(Y_{0q}^{S_z}) \tag{18}$$

where $R(y)$ is the function defined by:

$$R(y) = \int_0^\infty \frac{1}{\exp(\omega-y)+1} d\omega = \ln(1+\exp y) \tag{19}$$

The function R is an increasing function of its argument and for large values becomes approximately proportional to it in good agreement, according to Eq.(18), with the phenomenological assumption of the proportionality to $X_{0q}^{S_z}$ in [2]! At large negative y it becomes proportional to $\exp(-y)$, which corresponds to the Boltzman limit, while the previous one corresponds to the high degeneracy limit for the Fermi gas.

The factors:

$$R(-kX_{0q}^{S_z}) \simeq \exp(-kX_{0q}^{S_z}) \tag{20}$$

are different from the factors $\frac{1}{X_{0q}^{S_z}}$ empirically introduced in [2]. To get the same ratio for $\frac{\bar{u}(x\simeq0)}{d(x\simeq0)}$, in such a way to reproduce the agreement with the low x E866 data [5] one should take $k \simeq 4$, while the similarity of the non-diffractive light quark x distributions introduced in [2] and of the ones found after the p_T^2 integration with an appropriate choice of $f_q(x)$ increases with k. By comparing with the same set of very precise data cosidered in [2], one can look for the most suitable value for k [7].

In conclusion the extension to the transverse degrees of freedom and the sum rule written in eq.(13) have improved the theoretical reliability of the phenomenologically successful quantum statistical approach developed by us in the last decade [8] [2].

References

1. J. Soffer, These Proceedings
2. C. Bourrely, F. Buccella and J. Soffer, Eur. Phys. J. C **23**, 487 (2002)
3. R.S. Bhalerao, Phys. Rev. C **C63**, 025208 (2001);
4. C. Bourrely, F. Buccella and J. Soffer, Eur. Phys. J. C41, 327 (2005)
5. FNAL Nusea Collaboration, E. A. Hawker *et al.*, Phys. Rev. Lett. **80**, 3715 (1998); J. C. Peng *et al.*, Phys. Rev. D **58**, 092004 (1998); R. S. Towell *et al.*, Phys. Rev. D **64**, 052002 (2001); J.C. Webb *et al.*, submitted to Phys. Rev. Lett. [hep-ex/0302019]
6. F. Buccella and L. Popova, Mod. Phys. Lett. A **17**, 2627 (2002)
7. C. Bourrely, F. Buccella and J. Soffer, submitted to Phys. Lett.B [hep-th/0507328]
8. C. Bourrely, F. Buccella, G. Miele, G. Migliore, J. Soffer and V. Tibullo, Z. Phys. C **62**, 431 (1994)

RESEARCH PERSPECTIVES AT JEFFERSON LAB*

KEES DE JAGER

Jefferson Laboratory, Newport News, VA 23606, USA

The plans for upgrading the CEBAF accelerator at Jefferson Lab to 12 GeV are presented. The research program supporting that upgrade is illustrated with a few selected examples. The instrumentation under design to carry out that research program is discussed.

1. Introduction

The design parameters of the Continuous Electron Beam Accelerator Facility (CEBAF) at the Thomas Jefferson National Accelerator Facility (JLab) were defined over two decades ago. Since then our understanding of the behaviour of strongly interacting matter has evolved significantly, providing important new classes of experimental questions which can be optimally adressed by a CEBAF-type accelerator at higher energy. The original design of the facility, coupled to developments in superconducting RF technology, makes it feasible to triple the initial design value of CEBAF's beam energy to 12 GeV in a cost-effective manner.

The research program with the 12 GeV upgrade will provide breakthroughs in two key areas: (1) mapping gluonic excitations of mesons and understanding the origin of quark confinement and (2) searches for physics beyond the Standard Model. The upgrade will also provide important advances in two additional areas: (3) a direct exploration of the quark-gluon structure of the nucleon and (4) the physics of nuclei to understand the QCD basis for the nucleon-nucleon force and how nucleons and mesons arise as an approximation to the underlying quark-gluon structure. An overview of the upgrade research program is given in its Conceptual Design Report[1].

---

*This work was supported by DOE contract DE-AC05-84ER40150 Modification No. M175, under which the Southeastern Universities Research Association (SURA) operates the Thomas Jefferson National Accelerator Facility.

Lattice QCD calculations[2] have convincingly illustrated the linear quark-quark potential necessary for confinement. In a meson the quark and anti-quark are sources of color electric flux, which is trapped in a flux tube connecting the q and $\bar{q}$. However, very little is still known about the direct excitation of that flux tube. The observation of such direct manifestations of gluonic degrees of freedom will provide understanding of confinement[3]. The quantum numbers of the flux tube, added to those of a $q\bar{q}$ meson, can produce exotic hybrids with unique J^{PC} quantum numbers. These excitations can be probed far more effectively with photons than with π- or K-mesons, because the quark spins are aligned in the virtual vector-meson component of the photon. For a full partial-wave analysis of such excitations linearly polarized photons are a requisite. The Hall D research program will be focused on a definitive measurement of the spectrum of exotic hybrid mesons, expected in a mass range from 1 to 2.5 GeV/c^2.

One of the more compelling new opportunities with the 12 GeV upgrade will be a highly accurate measurement of the weak charge of the electron, via the parity-violating asymmetry in electron-electron (Møller) scattering. The achievable accuracy of such a measurement provides sensitivity to electron substructure to a scale of nearly 30 TeV. The measurement is also sensitive to the existence of new neutral gauge bosons in the range of 1 to 2 TeV; such model-dependent limits are comparable to those to be achieved by measurements at the Large Hadron Collider. Furthermore, the measurement will severely constrain the viability of SUSY models which violate R-parity. The upgraded beam energy will also make possible measurements of parity violation in deep inelastic scattering (PVDIS). On an isoscalar target at moderate x PVDIS is also sensitive to $\sin^2(\theta_W)$. A measurement of the PVDIS asymmetry will provide a very sensitive test of the electro-weak theory. Examples of additional PVDIS measurements are the value of $d(x)/u(x)$ as $x \to 1$, the search for evidence of charge symmetry violation at the partonic level, and the characterization of novel higher-twist effects. The PVDIS program will require the use of a new large-acceptance spectrometer/detector package.

A main focus of the research program will be the Generalized Parton Distributions (GPD) through the study of exclusive processes at large momentum transfer. The GPDs can be considered as overlap integrals between different components of the hadronic wave function[4], governed by the selection of the final state. Measurements of these GPDs will thus make it possible to map out quark and gluon wave functions. The orbital angular momentum contribution to the nucleon spin can be directly accessed through GPDs. Factorization is an essential ingredient in the extraction of GPDs. For Deeply Virtual Compton Scattering (DVCS) scaling will have been achieved at 11 GeV, but this has to be established experimentally for other processes.

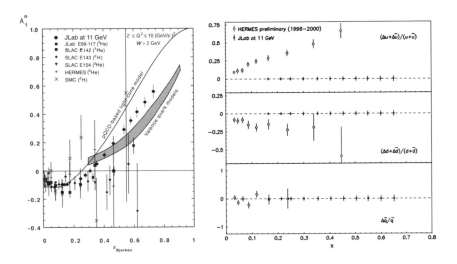

Figure 1. On the left is shown the projected measurement of A_1^n, on the right the projected determination of various combinations of polarized valence and sea quark distributions from semi-inclusive deep inelastic scattering.

One of the most fundamental properties of the nucleon is the structure of its quark distributions at higher x-values, where the physics of the valence quarks is cleanly exposed. The 12 GeV Upgrade will for the first time (by providing the necessary combination of high beam intensity and reach in Q^2) allow to map out the valence quark distributions at large x with high precision. Figure 1 (left) shows how the neutron polarization asymmetry A_1^n can be measured up to x-values close to 0.8 outside the nucleon resonance region. Most dynamical models predict that in the limit where a single valence up or down quark carries all of the nucleons momentum ($x \to 1$), it will also carry all of the spin polarization. Existing data on A_1^n show no sign of making the predicted dramatic transition $A_1^n \to 1$ (recent data from Hall A show the first hint of a possible upturn at the largest x-value).

There is a similar lack of data on other deep inelastic scattering observables in this region. One example is the ratio of down to up quarks in the proton, $d(x)/u(x)$, whose large-x behavior is intimately related to the fact that the proton and neutron are the stable building blocks of nuclei. This ratio requires measurement of the neutron as well as the proton structure function. Information about the neutron has to be extracted from deuterium data, and is difficult to disentangle from nuclear effects at large x. Figure 2 shows the precision with which this fundamental ratio can be measured with the 12 GeV Upgrade. The proposed experiment will utilize a novel technique; detection of the slowly recoiling proton spectator will tag scattering events on a nearly on-shell neutron in a deuteron target. An independent measurement of $d(x)/u(x)$ can be made by exploiting the mirror symmetry of $A = 3$ nuclei in simultaneous measurements with ^{3}He and ^{3}H targets. Both methods are designed to largely eliminate the nuclear corrections, thereby permitting the d/u ratio to be extracted with unprecedented precision.

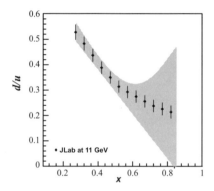

Figure 2. Projected measurement of the ratio of d- and u-quark momentum distributions, $d(x)/u(x)$, at large x. The shaded band represents the uncertainty in existing measurements due to nuclear Fermi motion effects.

The precise way in which the spin of the nucleon is distributed among its quark and gluon constituents is one of the most fundamental questions that can be addressed in nonperturbative QCD. Most of the experiments so far have focused on measuring the total quark and gluon contribution to the nucleon spin in inclusive deepinelastic scattering. In recent years the focus has moved to the investigation of specific aspects of the nucleon spin, such as the flavor asymmetries of sea quark distributions and quark transverse spin (transversity) distributions. The mapping of the flavor dependence of polarized valence and sea quark distributions and the determination of the quark transversity distributions require semi-inclusive measurements, in which the detected final-state hadron reveals information about the spin, flavor, and charge of the struck quark participating in the deep-inelastic process. The 12 GeV Upgrade will provide a unique opportunity to perform semi-inclusive measurements with high precision over a wide kinematic range, producing a detailed picture of the spin structure of the nucleon. Figure 1 (right) shows how polarized valence and sea quark distributions can be extracted from semi-inclusive deep inelastic scattering by detecting the leading $\pi^{\pm}$ and $K^{\pm}$ hadrons.

At 12 GeV, the details of the nucleon-nucleon force can be probed at distance scales much less than the pion Compton wave length, where the effects of two-pion exchange, vector-meson exchange, and quark exchange all compete. Although well constrained phenomenologically by the large body of pp and np elastic scattering data, it is not yet understood under what circumstances the effective nuclear force can be described in terms of the exchange of mesons, and when it is more efficient to describe the force in terms of the underlying quark-gluon exchange forces. Alternatively, the atomic nucleus can be used as a laboratory to study how the underlying QCD non-Abelian degrees of freedom manifest themselves. The idea is here to strike a quark inside the nucleus with such velocity that one can uniquely witness how hadrons emerge on their path through the nucleus. Our present sketchy understanding of this process will be vastly improved with the 12-GeV program at JLab.

2. Accclcrator

At present CEBAF accelerates electrons to 6 GeV by recirculating the beam four times through two superconducting linacs, each producing an energy gain of 600 MeV per pass. Both linac tunnels provide sufficient space to install five additional newly designed cryomodules. The new cryomodules will each provide over 100 MV (compared to the 28 MV from the existing ones), by increasing the gradient to 20 MV/m and the number of cavity cells from five to seven. This will result in a maximum energy gain per pass of 2.2 GeV, providing a maximum beam energy to Halls A, B and C of 11 GeV. The new Hall D will be provided with the desired maximum energy of 12 GeV by adding a tenth arc and recirculating the beam a fifth time through one linac. A total of 90 μA of CW beam can be provided at the maximum beam energy. Further modifications required are changing the dipoles in the arcs from C-type to H-type magnets, replacing a large number of power supplies and doubling the central helium liquifier capacity to 10 kW. An overview of the accelerator upgrade is shown in Fig. 3.

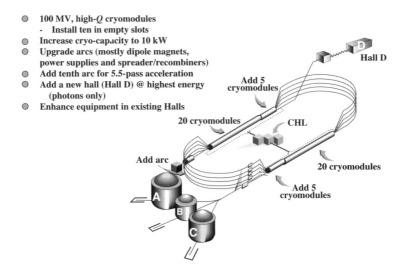

- 100 MV, high-Q cryomodules
  - Install ten in empty slots
- Increase cryo-capacity to 10 kW
- Upgrade arcs (mostly dipole magnets, power supplies and spreader/recombiners)
- Add tenth arc for 5.5-pass acceleration
- Add a new hall (Hall D) @ highest energy (photons only)
- Enhance equipment in existing Halls

Figure 3.   Overview of the accelerator upgrade to 12 GeV.

3. Hall Upgrades

The CEBAF Large Acceptance Spectrometer (CLAS) in Hall B is used for experiments that require the detection of several, loosely correlated particles in the hadronic final state at a limited luminosity. The CLAS12 detector has evolved from CLAS to meet the basic requirements for the study of the structure of nucleons and nuclei with the CEBAF 12 GeV upgrade. The main features are: 1) an operating luminosity of $L > 10^{35}$ $cm^{-2}s^{-1}$ for hydrogen targets, a ten-fold increase over current CLAS operating conditions; 2) detection capabilities and particle identification for forward-going high momentum charged and neutral particles; 3) improved hermeticity for the detection of charged particles and photons in regions where CLAS currently has no detection capabilities. CLAS12 makes use of several existing detector components. Major new components include new superconducting torus coils that cover only the forward-angle range, a new gas Čerenkov counter for pion identification, additions to the electromagnetic calorimeters, and the central detector.

The Hall C facility has generally been used for experiments which require high luminosity at moderate resolution. The core spectrometers are the High Momentum Spectrometer (HMS) and the Short Orbit Spectrometer (SOS). The HMS has a maximum momentum of 7.6 GeV/c. At a 12-GeV Jefferson Lab, Hall C will provide a new magnetic spectrometer, the Super High Momentum Spectrometer (SHMS), powerful enough to analyze charged particles with momenta approaching that of the highest energy beam. Charged particles with such high momenta are boosted by relativistic kinematics into the forward detection hemisphere. Therefore, the SHMS is designed to achieve angles down to 5.5°, and up to 25°. The SHMS will cover a solid angle up to 4 msr, and boasts a large momentum and target acceptance. The magnetic spectrometer pair will be rigidly connected to a central pivot.

The present base instrumentation in Hall A has been used for experiments which require high luminosity and high resolution in momentum and/or angle of at least one of the reaction products. The central elements are the two High Resolution Spectrometers (HRS). Both of these devices provide a momentum resolution of better than 2×10^{-4} and an angular resolution of better than 1 mrad with a design maximum central momentum of 4 GeV/c. The beamline into Hall A will be upgraded so that the hall will be able to accept the full range of beam energies available for two major purposes. The first will be to continue the use of the high resolution spectrometer pair. The second purpose for Hall A will be to stage major installation experiments. With a diameter of over 50 m, Hall A is the largest experimental hall at Jefferson Lab and can easily accommodate major installations such as the proposed parity-violation setups.

3.1. *Hall D*

The GlueX experiment will be housed in a new aboveground experimental hall (Hall D) located at the east end of the CEBAF north linac. A collimated beam of linearly polarized photons (with 40% polarization) of energy 8.5 to 9 GeV, optimum for production of exotic hybrids in its expected mass range, will be produced via coherent bremsstrahlung with 12 GeV electrons. This requires thin diamond crystal radiators. The scattered electron from the bremsstrahlung will be tagged with sufficient precision to determine the photon energy to within 0.1%.

The GlueX detector uses an existing 2.25 T superconducting solenoid that is currently being refurbished. An existing 3000-element lead-glass electromagnetic calorimeter will be reconfigured to match the downstream aperture of the solenoid. Inside the full length of the solenoid, a lead and scintillating fiber electromagnetic calorimeter will provide position and energy measurement for photons and TOF information for charged particles. A simple start counter will surround the 30 cm long liquid hydrogen target. This in turn will be surrounded by cylindrical straw-tube drift-chambers which will fill the region between the target and the cylindrical calorimeter. Planar drift chambers will be placed inside the solenoid downstream of the target to provide accurate track reconstruction for charged particles going in the forward direction.

This detector configuration has 4π hermeticity and momentum/energy and position information for charged particles and photons produced from incoming 9 GeV photons. It has been carefully optimized to carry out partial wave analysis of many-particle final states. The final planned photon flux is 10^8 photons/s. At this flux the experiment will acumulate in one year of running a factor of 100 more meson data than are presently available even from pion production.

4. Summary

In April of 2004 the US Department of Energy (DOE) signed CD-0 approval for the 12 GeV Upgrade project. With this Critical Decision #0 DOE acknowledges the mission need for this project. In April of 2005 the DOE Office of Science conducted a Science Review of the 12 GeV upgrade, which found the proposed research to have high scientific merit. Then, in July of 2005 DOE conducted a review of all aspects of the project's conceptual design. The review concluded that all of the requirements for CD- approval have been completed. Two further review processes in increasing level of detail, spaced 12 to 18 months apart, have to be successfully passed before funding for the project will be allocated.

References

1. *Conceptual Design Report for the Science and Experimental Equipment for the 12 GeV Upgrade of CEBAF*, 2005, http://www.jlab.org/div_dept /physics_division/GeV/doe_review/CDR_for_Science_Review.pdf.
2. G.S. Bali *et al.*, Proceedings of Int. Conf. on Quark Confinement and the Hadron spectrum, World Scientific, 1995, p. 225.
3. N. Isgur, R. Kokoski and J. Paton, Phys. Rev. Lett. 54 (1985) 869;
   S. Godfrey and J. Napolitano, Rev. Mod. Phys. 71 (1999) 1411.
4. X. Ji, Phys. Rev. Lett. 78 (1997) 610 ; A. Radyushkin, Phys. Lett. B 380 (1996) 417.

PERSPECTIVES WITH P̄ANDA

P. GIANOTTI

INFN-Laboratori Nazionali di Frascati
Via E. Fermi 40,
00044 Frascati - Italy
E-mail: paola.gianotti@lnf.infn.it

A new hadron facility is under construction at Darmstadt: FAIR (Facility for Antiproton and Ion Research). This will upgrade the scientific opportunities available at the GSI laboratory with new radioactive and relativistic ion's beams and a new antiproton's machine. Some aspects of the scientific program accessible with the new antiproton beam, that will be addressed by the P̄ANDA experiment, are here illustrated.

1. Introduction

The GSI laboratory in Germany is undergoing a major upgrade of the existing accelerator complex [1]. This upgrade foresees ion's beams of higher intensity and better quality, and, first for GSI, an antiproton's beam. At FAIR the antiprotons will be produced at the rate of $2 \times 10^7/s$ and then, after accumulation and cooling, will be transferred inside the High Energy Storage Ring (HESR) for the experimental activity. The HESR machine will be equipped with an internal target surrounded by a general purpose detector: P̄ANDA (Antiproton Annihilation at Darmstadt). The momentum of the antiprotons will vary between 1.5 and 15 GeV/c so that the maximum center-of-mass energy will reach 5.5 GeV, enough for the associate production of Ω_c which is the upper limit for the mass range of charmonium hybrids predictions.

P̄ANDA will perform a complete program of hadron spectroscopy to test many aspects of Quantum Chromo Dynamics (QCD), the generally accepted theory of strong interaction. The aim is to investigate both the dynamic of the interaction of fundamental particles, and the existence of new forms of matter: particles with glue content (glueballs, hybrids), extra charmonium states, nuclei with an explicit strange quark content, particles produced inside the nuclear medium. The importance of these studies is related to our capability of predicting, confirming, and explaining the physical states of the theory, that, in other words, means an exhaustive understanding of the strong interaction mechanism.

In the past, experiments with antiprotons have proven to be a rich source of information in this field, and with the new GSI antiproton's machine all the above mentioned topics will be addressed with a more powerful detector and dedicated experimental campaigns. $\overline{P}$ANDA will be the new general purpose detector optimized to accomplish this complete hadron physics program. In the following, the main topics of this program will be illustrated.

2. Charmonium Spectroscopy

The first charmonium state has been discovered in 1974, but there are still many open questions regarding the properties of the systems above and below the open charm threshold. The goal of $\overline{P}$ANDA is to make comprehensive measurements of all the charmonium states via their electromagnetic and hadronic decays. Furthermore, states with explicit glue content (hybrids and glueballs) are also expected, and could be detected in this energy region more easily than at lower mass. $\bar{p}p$ formation experiments will generate charmonium and charmonium exotics with ordinary quantum numbers with high cross sections, while production experiments could also yield states with non $\bar{q}q$ quantum numbers. Above the $D\overline{D}$ threshold very little is known. Recently, an important discovery in this region as been performed by the Belle collaboration: a new state named X(3872) [2] has been seen in the $\pi^+\pi^-J/\psi$ mass spectrum. The nature of this new, narrow, state is still controversial: speculation ranges from a $D^0\overline{D^{0*}}$ molecule to a $^3D_2$ state. From the theoretical point of view all these interpretations are unsatisfactory and further measurements of other decay channels are necessary. This new measurements will be ideally performed in $\bar{p}p$ formation experiments.

3. Hadrons in nuclear matter

The investigation of hadron's properties when embedded in nuclear matter is of paramount importance to understand the origin of hadron masses. The breaking of the chiral symmetry in QCD is influencing the mass generation mechanism, therefore this can be studied into an environment (the nucleus) where this symmetry is partially restored. So far, the experimental investigations have been focused on the light quark sector; the in-medium potential of pions has been deduced from spectroscopic information obtained in the study of deeply-bound pionic states [3]. Measurements of K production in proton-nucleus collisions[4] and heavy-ion collisions [5] indicate a repulsive and an attractive mass shift for K^+ and K^-, respectively, in nuclear matter. The high intensity HESR antiproton beam allow to extend this research to the charmed mesons. Up to now very little is known about charm in nuclear matter and theoretical predictions are widely model dependent. On the other hand the precise knowledge of J/ψ production cross sections in $\bar{p}$ annihilation on a series of different nuclei is required also by heavy ion collision experiments where a signature of quark-gluon-plasma formation is the suppression of charm production. On the other hand D mesons provide the unique opportunity to study the in-medium dynamics of a system with a single light quark. From the experimental point of view, the in-medium mass of charmonium states can be determined using their leptonic decay modes, whereas the D and $\overline{D}$ mesons can be identified via their hadronic decays into $\overline{K}$ and K mesons.

4. Double Λ-hypernuclei

Double Λ-hypernuclei are a unique source of data on ΛΛ interaction. However, only 6 double Λ-hypernuclei are presently known. The use of the $\bar{p}$ beam, in conjunction with a suitably designed apparatus, will allow the $\overline{P}$ANDA detector to be transformed into a ΛΛ hypernuclear factory.

5. The $\overline{P}$ANDA detector

To accomplish the scientific program illustrated in the previous sections, an ideal detector must cover a nearly full solid angle, it must stand high rates of events $(2 \times 10^7$ ann./s) having good capabilities of particle's identification and momentum resolution. Additional requirements are: excellent primary and secondary vertex reconstruction, modularity to allow different configurations to perform the different measurements.

In order to fulfill all these requirements, the $\overline{P}$ANDA collaboration is designing an apparatus consisting of two components: a Target Spectrometer (TS) surrounding the interaction region, and a Forward Spectrometer (FS) to close the acceptance in the forward region. The main elements of the TS, located inside a superconducting solenoid with a diameter of 1.6 m and a length of 2.5 m, are: a silicon vertex detector, a charged particles tracking system consisting of a barrel of straw tubes and Mini Drift Chambers (MDC), a cylindrical plus a planar DIRC Cherenkov for particle's identification, an electromagnetic calorimeter of PbWO$_4$ crystals. Outside the iron yoke muon detectors will be installed. Downstream a second conventional dipole magnet will be used to determine the momentum of the particles emitted within 0 and 5° in the vertical plane by the FS. It will consist of MDCs for charged particles tracking, TOF walls, RICH counters plus an electromagnetic and an hadronic calorimeter followed by a set of muon chambers. A sketch of the proposed detector is shown in figure 1.

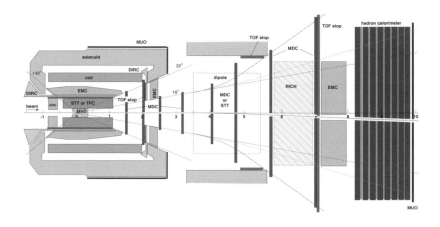

Figure 1.   Top view of the $\overline{\text{P}}$ANDA spectrometer.

6. Conclusions

A rich and unique physics program, with emphasis on charmed particles will be investigated at the HESR at GSI allowing to shed new light on many unclear aspects of QCD. Antiproton induced reactions exhibit unique features:

- gluonic hadrons have high production rates in $p\bar{p}$ annihilation;
- high statistics data with low multiplicity events will be produced;
- many states can be directly formed and investigated

Furthermore, thanks to the versatility of the $\overline{\text{P}}$ANDA spectrometer, other topics like Drell-Yan reactions, CP violation studies, rare D meson decay modes, proton form factors measurement could be addressed.

References

1. *An international Accelerator Facility for Beams of Ions and Antiprotons.* Conceptual Design Report, November 2001. http://www.gsi.de/GSI-Future/cdr.
2. S. K. Choi *et al.*, *Phys. Rev. Lett.* **91**, 262002 (2003).
3. H. Geissel *et al.*, *Phys. Rev. Lett.* **88**, 122301 (2002); H. Geissel *et al.*, *Phys. Lett.* **B549**, 64 (2002); K. Suzuki *et al.*, *Phys. Rev. Lett.* **92** 072302 (2002).
4. M. Nekipelov *et al.*, *Phys. Lett.* **B540**, 207 (2002); Z. Rudy *et al.*, *Eur. Phys. J.* **A15**, 303 (2002).
5. Y. Shin *et al.*, *Phys. Rev. Lett.* **81** 1576 (1998); R. Barth *et al.*, *Phys. Rev. Lett.* **78** 4007 (1997); F. Laue *et al.*, *Phys. Rev. Lett.* **82** 1640 (1999)

TRANSVERSITY MEASUREMENT WITH POLARIZED PROTON AND ANTIPROTON INTERACTIONS AT GSI : THE PAX EXPERIMENT

P.F.DALPIAZ

*Physics Departement of the Ferrara University and INFN, Sezione di Ferrara
Via Saragat 1, 44100 Ferrara, Italy; E-mail: pfd@fe.infn.it*

It has recently been suggested by the PAX collaboration that collisions of transversely polarized protons and antiprotons at the GSI-FAIR can be used to determine the nucleon's transversity densities from measurements of the double-spin asymmetry for the Drell-Yan process. The theoretical expectations for this observable are in the 0.3–0.4 range at the FAIR-HESR enrgies. PAX therefore proposes to build a polarized antiproton stored beam suitable for this measurament. Polarized antiprotons will be produced by spin filtering with an internal polarized gas target in a storage ring. The design and performance of the accelerator setup, and of the the detector will be briefly outlined.

1. Transversity

The complete understanding of the partonic structure of a spin $\frac{1}{2}$ object like a nucleon at leading–twist level is given in terms of the unpolarized distribution functions $q(x, Q^2)$, the longitudinaly polarized, or helicity, distributions $\Delta q(x, Q^2)$, and by the transversity distributions $\delta q(x, Q^2)$ [a]. Combined experimental and theoretical efforts in deep inelastic scattering have led to an improved understanding of q and Δq distributions. In contrast, $\delta q(x, Q^2)$ remain quantities about which we have the least knowledge, even though there are now first indications [2] that some of them are non-vanishing.

The δq were first introduced in ref. [3]. They are defined as [4,3,5,1] the difference of probabilities for finding a parton of flavor q at scale Q^2 and light-cone momentum fraction x with its spin aligned ($\uparrow\uparrow$) or anti-aligned ($\downarrow\uparrow$) to that of the transversely polarized nucleon:

$$\delta q(x, Q^2) \equiv q_{\uparrow\uparrow}(x, Q^2) - q_{\downarrow\uparrow}(x, Q^2) .$$ (1)

---

[a] Various other notations for the distribution functions are currently used. See [1] for an extensive discussion on terminology.

By virtue of factorization theorems [6,7], the parton densities can be probed universally in a variety of inelastic scattering processes for which it is possible to separate ("factorize") the long-distance physics relating to nucleon structure from a partonic short-distance scattering that can be calculated in QCD perturbation theory. It was realized a long time ago [4,3,5] that due to its chirally-odd structure, transversity decouples from inclusive deeply-inelastic scattering, but that inelastic collisions of two transversely polarized nucleons should offer good possibilities to access transversity. In particular, the Drell-Yan processes $pp \to l^+l^-X$, $p\bar{p} \to l^+l^-X$ ($l = e, \mu$) were identified as promising sources of information on transversity [8,9,10]. There is no gluon transversity distribution at leading twist [4,5]. For the Drell-Yan process, the lowest-order partonic process is $q\bar{q} \to \gamma^*$, with gluonic contributions to the unpolarized cross section in the denominator of the transverse double-spin asymmetry

$$A_{TT} = \frac{\sigma^{\uparrow\uparrow} - \sigma^{\uparrow\downarrow}}{\sigma^{\uparrow\uparrow} + \sigma^{\uparrow\downarrow}} \tag{2}$$

only arising as higher-order corrections. Therefore, A_{TT} may be sizable for the Drell-Yan process, in contrast to other hadronic processes.

The PAX collaboration has recently proposed to add polarization to planned $\bar{p}p$ collision experiments at the GSI, and to perform measurements of A_{TT} for the Drell-Yan process [11]. Infact, unique information on transversity could be obtained in this way. First of all, in $\bar{p}p$ collisions the Drell-Yan process mainly probes products of two *quark* densities, $\delta q \times \delta q$, since the distribution of antiquarks in antiprotons equals that of quarks in the proton. In addition, kinematics in the proposed experiments are such that rather large partonic momentum fractions, $x \geq 0.1$, are probed. One therefore accesses the valence region of the nucleon. Estimates [12,13] for the GSI PAX experiment show that the expected spin asymmetry A_{TT} should be very large, and can reach 30% or even more.

Information on transversity should be gathered from polarized proton-proton collisions at the BNL Relativistic Heavy Ion Collider (RHIC). In pp collisions, however, the Drell-Yan process probes products of valence quark and sea antiquark distributions. It is possible that antiquarks in the nucleon carry only little transverse polarization since, due to the absence of a gluon transversity, a source for the perturbative generation of transversity sea quarks from $g \to q\bar{q}$ splitting is missing. In addition, at RHIC energies and for Drell-Yan masses of a few GeV, the partonic momentum fractions are fairly small, so that the denominator of A_{TT} is large due to the small-x

rise of the unpolarized sea quark distributions. Thus, even for the Drell-Yan process, the spin asymmetry A_{TT} at RHIC will probably be at most a few per cent, as theoretical studies have shown [9,10].

2. Antiproton polarization and HESR upgrade

For more than two decades, physicists have tried to produce beams of polarized antiprotons [14]. Conventional methods like atomic beam sources cannot be applied, since antiprotons annihilate with matter. Polarized antiprotons have been produced from the decay in flight of $\bar{\Lambda}$ hyperons at Fermilab[15] or scattering of antiprotons off a liquid hydrogen target [16]. Unfortunately, both approaches do not allow efficient accumulation in a storage ring, which would greatly enhance the luminosity. Spin splitting using the Stern–Gerlach separation of the given magnetic substates in a stored antiproton beam was proposed in 1985 [17]. Although the theoretical understanding has much improved since then [18], spin splitting using a stored beam has yet to be observed experimentally.

In 1992 an experiment at the Test Storage Ring (TSR) at MPI Heidelberg showed that an initially unpolarized stored 23 MeV proton beam can be polarized by spin–dependent interaction with a polarized hydrogen gas target [19,20,21]. In the presence of polarized protons of magnetic quantum number $m = \frac{1}{2}$ in the target, beam protons with $m = \frac{1}{2}$ are scattered less often, than those with $m = -\frac{1}{2}$, which eventually caused the stored beam to acquire a polarization parallel to the proton spin of the hydrogen atoms during spin filtering. A beam polarization of more then 2% in two hours has been clearly observed.

On this basis, the PAX collaboration has elaborated a viable scheme which would allow to reach a polarisation of the stored antiprotons at HESR FAIR of $\simeq 0.2 - 0.4$ in a dedicated low–energy Antiproton Polarizer Ring (APR), for which energy and acceptance have been optimized. The transfer of polarized antiprotons into the HESR requires pre–acceleration in a dedicated COSY–like booster ring (CSR, Cooler Synchrotron Ring), in which protons or antiprotons can be stored with accelerated up to 3.5 GeV/c and transferred to the HESR ring, that should be run as a synchrotron.

The overall machine setup at the HESR–FAIR is schematically depicted in Fig. 1. It may be noted that the CSR has a straight section, where a PAX detector could be installed, running parallel to a straight section of the HESR. By deflection of the HESR beam into the straight section of the

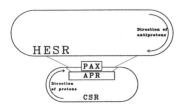

Figure 1. The proposed accelerator set–up at the HESR–FAIR

CSR, both a collider or a fixed–target mode become feasible at the PAX interaction point. The PAX detector has been designed to be used for fixed target and collider mode measuraments.[b] It is well–suited to provide large invariant-mass e^+e^- pair detection, from both Drell-Yan reactions and $\bar{p}p$ annihilations and to efficiently detect secondaries in two body reactions, like elastic scattering events. It has a large angle acceptance, is sensitive to electron pairs, with an adequate hadron rejection capacity. See for details [11]

3. Conclusion

The PAX Collaboration proposes the first ever direct measurement of the quark transversity distribution δq, by measuring the double transverse spin asymmetry A_{TT} in Drell–Yan processes $p^\uparrow\bar{p}^\uparrow \to e^+e^-X$. The reactions will be produced with a beam of polarized antiprotons from 1.5 GeV/c up to 15 GeV/c circulating in the HESR, colliding on a beam of polarized protons up to 3.5 GeV/c circulating in the CSR. By proper variation of the energy of the two colliding beams, this setup would allow a measurement of δq in the valence region of $x > 0.05$, with corresponding $4 < Q^2 < 100$ GeV^2 . A_{TT} is predicted to be larger than 20 % over the full kinematic range, till the highest reachable center-of-mass energy of $\sqrt{s} \sim \sqrt{200}$. With a luminosity of $2 \cdot 10^{30}$ cm$^{-2}s^{-1}$ about ~ 1000 events per day can be expected.

It is worthwhile to stress here that the PAX physics program adresses, among transversity, other very interesting items . See for reference [11] , [22]. Running in fixed target mode, a beam of unpolarized or polarized antiprotons with momentum up to 3.5 GeV/c in the CSR ring, will collide on a polarized hydrogen target in the PAX detector. This allows, for the first time, the measurement of the time–like proton form factors in single and

---

[b]For this pourpose the tracking magnet is a toroid magnet.

double polarized $\bar{p}p$ interactions from close to the threshold up to $Q^2 = 8.5$ GeV^2. It would be possible to determine several double spin asymmetries in elastic $\bar{p}^\uparrow p^\uparrow$ scattering. By detecting back scattered antiprotons one can also explore hard scattering regions of large t: in the proton–proton scattering reaching the same region of t requires twice higher energy. At present, there are no competing facilities at which all these topical issues can be addressed.

References

1. A comprehensive review on transversity can be found in: V. Barone, A. Drago and P. G. Ratcliffe, Phys. Rept. **359**, 1 (2002) [arXiv:hep-ph/0104283].
2. A. Airapetian et al. [HERMES Collaboration], Phys. Rev. Lett. **94**, 012002 (2005) [arXiv:hep-ex/0408013].
3. J.P. Ralston and D.E. Soper, Nucl. Phys. B **152**, 109 (1979).
4. R.L. Jaffe and X. Ji, Phys. Rev. Lett. **67**, 552 (1991); Nucl. Phys. B **375**, 527 (1992).
5. X. Artru and M. Mekhfi, Z. Phys. C **45**, 669 (1990).
6. J.C. Collins, D.E. Soper, and G. Sterman, Nucl. Phys. B **308**, 833 (1988);
7. J.C. Collins, Nucl. Phys. B **394**, 169 (1993) [arXiv:hep-ph/9207265].
8. J.L. Cortes et al., Z. Phys. C **55**, 409 (1992); V. Barone, et al. Phys. Rev. D **56**, 527 (1997) [arXiv:hep-ph/9702239];
9. O. Martin et al., Phys. Rev. D **57**, 3084 (1998) [arXiv:hep-ph/9710300].
10. O. Martin et al., Phys. Rev. D **60**, 117502 (1999) [arXiv:hep-ph/9902250].
11. P. Lenisa and F. Rathmann [the PAX Collaboration], arXiv:hep-ex/0505054 and http://www.fz-juelich.de/ikp/pax/;
12. M. Anselmino et al., Phys. Lett. B **594**, 97 (2004) [arXiv:hep-ph/0403114].
13. A. V. Efremov et al., Eur. Phys. J. C **35**, 207 (2004) [arXiv:hep-ph/0403124].
14. Proc. of the Workshop on Polarized Antiprotons, Bodega Bay, CA, 1985, Eds. A. D. Krisch, A. M. T. Lin, and O. Chamberlain, AIP Conf. Proc. **145** (AIP, New York, 1986).
15. D.P. Grosnick et al., Nucl. Instrum. Methods **A290** (1990) 269.
16. H. Spinka et al., Proc. of the 8th Int. Symp. on Polarization Phenomena in Nuclear Physics, Bloomington, Indiana, 1994, Eds. E.J. Stephenson and S.E. Vigdor, AIP Conf. Proc. **339** (AIP, Woodbury, NY, 1995), p. 713.
17. T.O. Niinikoski and R. Rossmanith, Nucl. Instrum. Methods **A255** (1987) 460.
18. P. Cameron et al., Proc. of the 15th Int. Spin Physics Symp., Upton, New York, 2002, Eds. Y.I. Makdisi, A.U. Luccio, and W.W. MacKay, AIP Conf. Proc. **675** (AIP, Melville, NY, 2003), p. 781.
19. F. Rathmann et al., Phys. Rev. Lett. **71** (1993) 1379.
20. K. Zapfe et al., Rev. Sci. Instrum. **66** (1995) 28.
21. K. Zapfe et al., Nucl. Instrum. Methods **A368** (1996) 627.
22. S.J. Brodsky et al., Phys. Rev. D **69**, 054022 (2004).

Conclusion

SUMMARY TALK ON QUARK-HADRON DUALITY*

PAUL HOYER

Department of Physical Sciences and Helsinki Institute of Physics
POB 64, FIN - 00014 University of Helsinki, Finland
and
NORDITA, Blegdamsvej 17, DK-2100 Copenhagen, Denmark
E-mail: paul.hoyer@helsinki.fi

I ascribe the origins of Bloom-Gilman duality in DIS to a separation of scales between the hard subprocess and soft resonance formation. The success of duality indicates that the subprocesses of exclusive form factors are the same as in DIS. The observed dominance of the longitudinal structure function at large x in $\pi N \to \mu^+\mu^- X$ can explain why local duality works for DIS with a pion target. The failure of duality in semi-exclusive processes indicates that high momentum transfer t is not sufficient to make the corresponding subprocesses compact.

This meeting[a] demonstrated the exciting progress made in the last few years on duality in Deep Inelastic Scattering (DIS). Building on the work of Bloom and Gilman[1] 35 years ago, high precision data principally from Jlab[2,3] and DESY[4] has reopened the field, allowing detailed studies of duality including spin dependence and nuclear effects. In this written summary I focus on just a few aspects of duality that I think give clues to the underlying QCD dynamics. I refer to the presentations given at the workshop for the many important results that I cannot cover here. The comprehensive review[5] by Melnitchouk, Ent and Keppel covers the experimental and theoretical results on duality available before the workshop.

1. Duality and the uncertainty relation

The duality between resonances and hard perturbative processes is most easily visualized in e^+e^- annihilations, where vector mesons average the

---

*Work partially supported by the Academy of Finland through grant 102046.
[a]First Workshop on Quark-Hadron Duality and the Transition to pQCD, Laboratori Nazionali di Frascati, June 6-8 2005; http://www.lnf.infn.it/conference/duality05/ .

asymptotic $e^+e^- \to Q\bar{Q}$ cross section[b]. Quarks with large mass M_Q are produced at a short time-scale $1/M_Q$. Resonances form at a longer time-scale $1/\Delta M_{Q\bar{Q}}$ characterized by their mass differences. Thus resonance formation is *incoherent with the hard subprocess*, *i.e.*, it cannot affect the quark production probability. At the time of resonance formation the total energy uncertainty $\Delta E \sim \Delta M_{Q\bar{Q}}$ limits the range within which the perturbative cross section may be "reshuffled" into resonance peaks.

The local duality between single resonances and the perturbative e^+e^- cross section can thus be viewed as a consequence of the relative softness of the interactions involved in resonance formation. In more differential quantities, such as spin-dependent structure functions, the scale ΔM refers to mass differences between resonances having the same spin dependence. Thus semilocal duality in DIS is found to work separately for the $N(940)$ and $\Delta(1232)$ in spin-averaged cross sections, whereas a broader averaging region appears to be required for spin-dependent quantities[5,7,8]. The larger ΔM may also explain why BG duality sets in at a higher value of Q^2 in the spin-dependent structure function, since the subprocess and resonance scales need to be clearly separated.

2. Duality in Deep Inelastic Scattering

Compared to e^+e^-, BG duality in DIS opens up a new dimension: At each value of Q^2 there is a whole range of Bjorken x-values in which the asymptotic cross section may be compared to the resonance contributions (Fig. 1). As Q^2 increases, a given resonance N^* of mass M contributes at an increasing value of $x = Q^2/(Q^2+M^2)$, thus "sliding" along the scaling curve $F_2(x)$ towards $x = 1$. If the transition form factor $F_{N \to N^*}(Q^2) \sim 1/Q^{2n}$ then local duality requires that the inclusive quark distribution behaves as $F_{q/N}(x) \sim (1-x)^{2n-1}$ for $x \to 1$.

In DIS the virtual photon scatters incoherently from each quark in the target. Hence the cross section depends on the sum of squared quark charges,

$$F_2 \sim \sum_q e_q^2 \tag{1}$$

Exclusive form factors (Fig. 2) are assumed[9] be built from target Fock states whose transverse dimensions are compatible with the photon resolution.

---

[b]This is, in particular, exploited in the "QCD Sum Rules"[6].

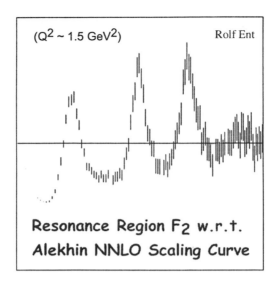

Figure 1. Comparison[7] between the x-dependence of the $F_2(x, Q^2)$ structure function measured at low $Q^2 \sim 1.5$ GeV^2 (errror bars with resonance structures) and the scaling curve measured at high Q^2 (horizontal line).

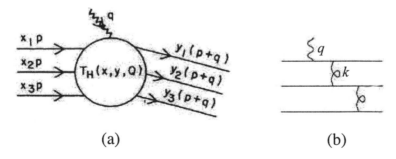

(a)   (b)

Figure 2. QCD form factor dynamics according to Lepage and Brodsky[9]. A generic diagram of the hard subprocess T_H in (a) is shown in (b). The momenta k of the gluon exchanges scale with the photon momentum q.

The gluons exchanged in the hard subprocess shown in Fig. 2(b) then have momenta $k \sim \mathcal{O}(Q)$, and the photon couples coherently to all quarks,

$$F_{N \to N^*} \sim \left(\sum_q e_q \right)^2 \qquad (2)$$

The electric charges of the quarks are unrelated to QCD dynamics. The different weighting in (1) and (2) thus appears incompatible with BG duality. The observation[10] that the interference terms in (2) cancel when averaging over resonances with different parity is not sufficient, since duality works within each resonance region (Fig. 1).

In the previous Section we saw that duality can be understood as a consequence of the unequal time scales involved in the hard subprocess and resonance formation. Once the hard process has "happened", later interactions can only redistribute the (inclusive) cross section within a limited mass range. Such an explanation requires that the hard subprocesses in resonance form factors and in the scaling DIS cross section *are the same.* In particular, the Fock states contributing to form factors must have a transverse size exceeding the photon resolution, so that coherent scattering from several quarks is suppressed.

PQCD calculations of proton and pion form factors[11] indicate that the virtuality k^2 of the gluons in Fig. 2(b) is typically much smaller than the photon virtuality Q^2. This means that the photon will effectively couple incoherently to the quarks, and the weighting of quark charges will be according to (1), just as in DIS.

The fact that BG duality works furthermore indicates that the hard subprocess has reached its scaling limit at the moderate Q^2 value of Fig. 1. If the resonances build the scaling distribution they must dominate the inclusive cross section at the corresponding value of x (given by the resonance mass). The $\Delta(1232)$ actually decreases faster with Q^2 than the scaling cross section – but the difference is taken up by "background"[12]. Thus BG duality still works locally in the $\Delta(1232)$ region, by relating the scaling (high Q^2) curve to the combined production of resonance and background at lower Q^2. This again requires that the subprocess has reached its scaling limit (in Q^2 at fixed x), whereas the $\Delta(1232)$/background ratio, which is determined by the soft hadronization process, decreases as $x \to 1$.

3. The longitudinal structure function

BG duality has been found to work[5,7] also in the longitudinal photon structure function $F_L(x, Q^2)$ for $Q^2 \gtrsim 1.5$ GeV^2. This is an important check, as subprocesses involving longitudinal photons differ from those of transverse photons (*e.g.*, longitudinal photons do not couple to on-shell spin $\frac{1}{2}$ quarks at leading twist). The success of duality again indicates that resonances are produced by the same hard subprocesses as those responsible for the scaling longitudinal structure function.

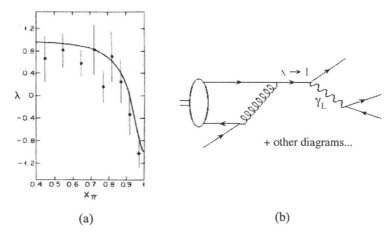

(a)   (b)

Figure 3. (a) The parameter λ of the angular distribution of the muons in $\pi N \rightarrow \mu^+\mu^- X$, as a function of the fractional momentum x_π of the projectile pion carried by the muon pair[13]. $\lambda = +1$ (-1) corresponds to a transversely (longitudinally) polarized virtual photon. The solid line is from Ref.[14]. (b) A generic diagram contributing in the limit $Q^2 \rightarrow \infty$ with $Q^2(1-x) = M^2$ held fixed. The $x \rightarrow 1$ quark propagator has virtuality of $\mathcal{O}\left(Q^2\right)$ and is thus coherent with the virtual photon.

The pion elastic form factor (measured by $e\pi \rightarrow e\pi$) gets a contribution only from longitudinal photon exchange. One would thus expect that the pion be dual to F_L measured on a pion target. While no DIS data on pion targets exist, the pion structure function has been measured in the Drell-Yan process $\pi N \rightarrow \mu^+\mu^- X$. The muon angular distribution shows[13] that the longitudinal structure function dominates at large x (Fig. 3a). This may be understood as a consequence of helicity conservation – the photon carries the helicity of the projectile as its momentum fraction $x_\pi \rightarrow 1$. Longitudinal photons can dominate over transverse photons in the limit $Q^2 \rightarrow \infty$, $x \rightarrow 1$ with $M^2 = Q^2(1-x)$ fixed since diagrams like the one in Fig. 3(b) contribute at leading order [14,15]. This limit is appropriate for the contribution of a given resonance of mass M to DIS. It is, however, quite different from the standard Bjorken limit where x is held fixed, the twist expansion is relevant and transverse photons dominate.

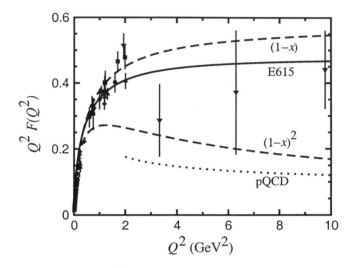

Figure 4. Duality comparison[17] of the elastic pion form factor (data points) with the E615 quark distribution of the pion measured in the Drell-Yan process.

The effective quark distribution in the pion was found[13] to behave as $f_{q/\pi}(x) \sim (1-x)^{1.12\pm0.18}$ at high x. In QCD one expects[16] $f_{q/\pi}(x) \sim (1-x)^2$ for transverse photons in the Bjorken limit. The data is not in conflict with QCD due to the contribution of longitudinal photons. The pion elastic form factor is expected[9] to behave as $F_\pi(Q^2) \sim 1/Q^2$ in QCD, which is consistent with the available data. Local BG duality for the pion then requires $f_{q/\pi}(x) \sim (1-x)^1$, tantalizingly close to the data. In fact, even the normalization is consistent with local duality for the pion[17], as shown in Fig. 4. This agreement lends further support to the dominance of the longitudinal structure function in the data at high x.

4. Extending duality: Semi-exclusive processes

I have argued that BG duality provides a tool for understanding the dominant dynamics of exclusive form factors. It is obviously important to try to generalize duality beyond DIS. Promising applications to semi-inclusive processes were already discussed at this workshop[7,18]. Here I shall mention a somewhat different approach which turns out not to work – but the failure is striking enough to be instructive.

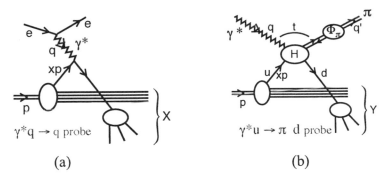

Figure 5. Similarity of dynamics in DIS (a) and semi-exclusive pion production (b). The hard subprocess $eq \to eq$ of DIS is replaced with the compact $\gamma^{(*)}u \to \pi^+ d$ subprocess in the semi-exclusive process.

As indicated in Fig. 5(a), in DIS we use the hard (PQCD calculable) subprocess $eq \to eq$ to probe the quark structure of a nucleon target. BG duality than implies that we can relate the inclusive cross sections measured at the same value of x but different Q^2 (and hence also different masses of the inclusive system X, down to the resonance region). Analogously, in Fig. 5(b) we are probing the quark distribution using a different subprocess $\gamma^{(*)}u \to \pi^+ d$. Insofar as this subprocess is hard, the quark pair forming the π^+ will be produced in a compact, color singlet configuration and will not further interact in the target due to color transparency. In a kinematical limit where the hadrons in the inclusive system Y are separated by a large rapidity gap from the π^+ we may calculate the cross section of this *semi-exclusive process* using PQCD and the standard parton distributions[19]. The soft dynamics forming the inclusive system Y is again incoherent with the hard subprocess and we may relate cross sections at various M_Y by appealing to duality.

An application of the above idea to $\gamma p \to \pi^+ n$ at large momentum transfer t gave an interestingly incorrect result[20]. There is no data on the semi-inclusive process $\gamma p \to \pi^+ Y$, but we expected to get a (ball-park) estimate for the exclusive process with $Y = n$ using local duality analogously to DIS. However, the calculation turned out to be off by nearly two orders of magnitude! The corresponding analysis of wide-angle Compton scattering $\gamma p \to \gamma p$ did not fare much better, underestimating the data by about a factor 10. To me the most likely explanation is that the subprocess H in Fig. 5(b) actually is soft even at high t (when the incoming photon is real). Clearly we have much to learn about the dynamics of processes involving large momentum transfers – and Bloom-Gilman duality provides us with a very powerful tool.

Acknowledgments

I am very grateful to the organizers for arranging this most topical workshop, and for asking me to give a contribution.

References

1. E. D. Bloom and F. J. Gilman, Phys. Rev. Lett. **25** (1970) 1140; Phys. Rev. D **4** (1971) 2901.
2. I. Niculescu *et al.*, Phys. Rev. Lett. **85** (2000) 1182.
3. I. Niculescu *et al.*, Phys. Rev. Lett. **85** (2000) 1186.
4. A. Airapetian *et al.* [HERMES Collaboration], Phys. Rev. Lett. **90** (2003) 092002 [arXiv:hep-ex/0209018].
5. W. Melnitchouk, R. Ent and C. Keppel, Phys. Rept. **406** (2005) 127 [arXiv:hep-ph/0501217].
6. M. A. Shifman, A. I. Vainshtein and V. I. Zakharov, Nucl. Phys. B **147** (1979) 385 and Nucl. Phys. B **147** (1979) 448.
7. R. Ent, These proceedings.
8. H. P. Blok, These proceedings.
9. G. P. Lepage and S. J. Brodsky, Phys. Rev. D **22** (1980) 2157.
10. F. E. Close and N. Isgur, Phys. Lett. B **509** (2001) 81 [arXiv:hep-ph/0102067].
11. J. Bolz, R. Jakob, P. Kroll, M. Bergmann and N. G. Stefanis, Z. Phys. C **66** (1995) 267 [arXiv:hep-ph/9405340]; B. Melic, B. Nizic and K. Passek, Phys. Rev. D **60** (1999) 074004 [arXiv:hep-ph/9802204].
12. C. E. Carlson and N. C. Mukhopadhyay, Phys. Rev. D **47** (1993) 1737.
13. S. Palestini *et al.*, Phys. Rev. Lett. **55** (1985) 2649.
14. E. L. Berger and S. J. Brodsky, Phys. Rev. Lett. **42** (1979) 940.
15. S. J. Brodsky, P. Hoyer, A. H. Mueller and W. K. Tang, Nucl. Phys. B **369** (1992) 519.
16. G. R. Farrar and D. R. Jackson, Phys. Rev. Lett. **35** (1975) 1416.
17. W. Melnitchouk, Eur. Phys. J. A **17** (2003) 223 [arXiv:hep-ph/0208258].
18. P. Bosted, These proceedings.
19. C. E. Carlson and A. B. Wakely, Phys. Rev. D **48** (1993) 2000; S. J. Brodsky, M. Diehl, P. Hoyer and S. Peigne, Phys. Lett. B **449** (1999) 306 [arXiv:hep-ph/9812277].
20. P. Eden, P. Hoyer and A. Khodjamirian, JHEP **0110** (2001) 040 [arXiv:hep-ph/0110297].

Appendix

Program

Monday, June 6
Tuesday, June 7
Wednesday, June 8

Morning **Duality in Photoproduction**
Chair: G. Pancheri (INFN Frascati)

09:00 "Duality in Vector Meson Production"
A. Donnachie (Manchester University)

09:30 "Pion Photoproduction from Nucleon and Scaling"
H. Gao (TUNL, Durham and Duke University)

10:00 "Onset of Scaling in Exclusive Processes"
M. Mirazita (INFN Frascati)

10:30 - Coffee break -
Chair: V. Muccifora (INFN Frascati)

11:00 "Exclusive baryon-antibaryon Production in $\gamma - \gamma$ at e^+e^- colliders"
T. Barillari (Max Planck Institute)

11:30 "Photoabsorption and Photoproduction on Nuclei in the Resonance Region"
S. Schadmand (IKP FZ Juelich)
Duality in Nuclei

12:00 "Nuclear Duality and Related Topics"
C. Keppel (Hampton University and JLAB)

Afternoon **Chair: N. Bianchi (INFN Frascati)**

14:00 "Quark-Hadron Duality of Hadronization in Nuclei"
X.N. Wang (BNL)

14:30 "Hadron Attenuation by (pre-)hadronic Final State Interaction at HERMES"
K. Gallmeister (Giessen University)

15:00 "Quest for the Quark Gluon Plasma: the High-Energy Frontier"
P. Giubellino (INFN Torino)

15:30 "Quark Gluon Plasma and Hadronic Gas on the Lattice"
M.P. Lombardo (INFN Frascati)

16:00 - Coffee break -
Duality in Neutrino Experiments
Chair: C. Keppel (Hampton University and JLAB)

16:30 "Neutrino and Electron Cross Sections at low Q^2"
A. Bodek (Rochester University)

17:00 "Neutrini Local Duality and Charge Symmetry Violation"
F. Steffens (Mackenzie University)
Duality and QCD

17:30 "Higher Twist Effects in Polarized DIS"
D. Stamenov (INR, Sofya)

18:00 "Higher Twist Effects in Polarized Experiments"
J.P. Chen (JLAB) for N. Liyanage (University of Virginia)

18:20 "Higher Twist Effect in Fragmentation Sector and its Experimental Extraction"
G. Xu (University of Houston)

Morning **Duality and QCD**
Chair: C. Carlson (College of William and Mary)

09:00 "Status of Unpolarised and Polarised Parton Distributions"
J. Blumlein (DESY Zeuthen)

09:30 "The transition from pQCD to npQCD"
A. Fantoni (INFN Frascati)

10:00 "Quark-Hadron Duality and High Excitations"
M. Shifman (Minnesota University)

10:30 - Coffee break -

11:00 "Hidden QCD Scales and Diquark Correlations"
A. Vainshtein (Minnesota University)

11:30 "Nearly-Conformal QCD and AdS/CFT"
G. de Teramond (Costa Rica University)

12:00 Matching Meson Resonances to OPE in QCD
A. Andrianov (INFN Bologna)

Afternoon **Future Perspectives**
Chair: A. Thoma (JLAB)

14:00 "Review of Generalised Parton Distributions"
M. Guidal (IPN Orsay)

14:30 "Experimental Overview on Exclusive Processes"
D. Hasch (INFN Frascati)

15:00 Transverse Polarization and Quark-Hadron Duality
O. Teryaev (BLTP, JINR)

15:30 "Extension to the Transverse Degree of Freedom of the Statistical Parton Distributions"
F. Buccella (Naples University and INFN)

16:00 - Coffee break -

Chair: P. Di Nezza

16:30 "Research Perspectives with the Jefferson Laboratory's 12 GeV Upgrade"
K. de Jager (JLAB)

17:10 "Perspectives with PANDA"
P. Gianotti (INFN Frascati)

17:30 "Transversity Measurement with Polarized antiproton-proton Interactions: the PAX Experiment"
P. Ferretti Dal Piaz (Ferrara University and INFN)

Conclusion

17:50 "Summary Talk"
P. Hoyer (Helsinky University)

AUTHOR INDEX

List of Participants

Alexander Andrianov
INFN Sezione di Bologna
Alexandr.Andrianov@bo.infn.it

Vladimir Andrianov
St. Petersburg Sate University and NIIF
andrian@heps.phys.spbu.ru

Teresa Barillari
Max Planck Institut fur Physik
barilla@mppmu.mpg.de

Nicola Bianchi
INFN Laboratori Nazionali di Frascati
bianchi@lnf.infn.it

Henk P. Blok
Vrije Universiteit and NIKHEF
henkb@nikhef.nl

Johannes Bluemlein
DESY
Johannes.Bluemlein@desy.de

Arie Bodek
University of Rochester
bodek@pas.rochester.edu

Peter Bosted
Jefferson Laboratory
bosted@jlab.org

Franco Buccella
Naples University and INFN Sezione di Napoli
buccella@na.infn.it

Mario Calvetti
INFN Laboratori Nazionali di Frascati
calvetti@lnf.infn.it

Carl Carlson
College of William and Mary
carlson@physics.wm.edu

Jian-Ping Chen
Jefferson Laboratory
jpchen@jlab.org

Eric Christy
Hampton University
christy@jlab.org

Donald Crabb
University of Virginia
dgc3q@unix.mail.virginia.edu

Paola Dal Piaz
Ferrara University and INFN Sezione di Ferrara
pfd@fe.infn.it

Kees de Jager
Jefferson Laboratory
kees@jlab.org

Enzo De Sanctis
INFN Laboratori Nazionali di Frascati
desanctis@lnf.infn.it

Guy de Teramond
University of Costa Rica
gdt@asterix.crnet.cr

Alexander Deur
Jefferson Laboratory
Alexandre.Deur@jlab.org

Pasquale Di Nezza
INFN Laboratori Nazionali di Frascati
dinezza@lnf.infn.it

Alexander Donnachie
University of Manchester
ad@a35.ph.man.ac.uk

Rolf Ent
Jefferson Laboratory
ent@jlab.org

Alessandra Fantoni
INFN Laboratori Nazionali di Frascati
fantoni@lnf.infn.it

Kai Gallmeister
University of Giessen
Kai.Gallmeister@theo.physik.uni-giessen.de

Haiyan Gao
TUNL, Durham and Duke University
gao@tunl.duke.edu

Paola Gianotti
INFN Laboratori Nazionali di Frascati
gianotti@lnf.infn.it

Paolo Giubellino
INFN Sezione di Torino
giubell@to.infn.it

Michel Guidal
IPN Orsay
guidal@ipno.in2p3.fr

Delia Hasch
INFN Laboratori Nazionali di Frascati

hasch@lnf.infn.it

Paul Hoyer
University of Helsinky
paul.hoyer@helsinki.fi

Xiaodong Jiang
Rutgers University
jiang@jlab.org

Mark Jones
Jefferson Laboratory
jones@jlab.org

Cynthia Keppel
Hampton University and Jefferson Laboratory
keppel@jlab.org

Alinaghi Khorramian
IPM
khorramiana@theory.ipm.ac.ir

Andrey Koshelkin
MEPHY

Simonetta Liuti
University of Virginia
sl4y@galileo.phys.virginia.edu

Nilanga Liyanage
University of Virginia
nilanga@virginia.edu

Maria Paola Lombardo
INFN Laboratori Nazionali di Frascati
lombardo@lnf.infn.it

Wally Melnitchouk
Jefferson Laboratory

wmelnitc@jlab.org

Zein-Eddine Meziani
Temple University
meziani@jlab.org

Marco Mirazita
INFN Laboratori Nazionali di Frascati
mirazita@lnf.infn.it

Valeria Muccifora
INFN Laboratori Nazionali di Frascati
muccifora@lnf.infn.it

Vitaly Okorokov
Moscow Engineering Physics Institute
okorokov@bnl.gov

Giulia Pancheri
INFN Laboratori Nazionali di Frascati
pancheri@lnf.infn.it

Donatella Pierluigi
INFN Laboratori Nazionali di Frascati
donatella.pierluigi@lnf.infn.it

Yelena Prok
Massachusetts Institute of Technology
yprok@jlab.org

Federico Ronchetti
INFN Laboratori Nazionali di Frascati
ronchetti@lnf.infn.it

Patrizia Rossi
INFN Laboratori Nazionali di Frascati
rossi@lnf.infn.it

Elena Santopinto

Genua University and INFN Sezione di Genova
elena.santopinto@ge.infn.it

Susan Schadmand
IKP FZ Juelich
s.schadmand@fz-juelich.de

Mikhail Shifman
University of Minnesota
shifman@umn.edu

Jacques Soffer
CNRS Marseille
Jacques.Soffer@cpt.univ-mrs.fr

Patricia Solvignon
Temple University
solvigno@jlab.org

Dimiter Stamenov
INR Sofya
stamenov@inrne.bas.bg

Fernando Steffens
University of Mackenzie
fsteffen@ift.unesp.br

Oleg Teryaev
BLTP, JINR
teryaev@theor.jinr.ru

Anthony Thomas
Jefferson Laboratory
awthomas@jlab.org

Howard Trottier
Simon Fraser University
trottier@sfu.ca

284

Arkady Vainshtein
University of Minnesota
vainshte@umn.edu

Laura Volpes
INFN Laboratori Nazionali di Frascati
good_riddance@virgilio.it

Xin-Nian Wang
Lawrence Berkeley National Laboratory
xnwang@lbl.gov

Guangua Xu
University of Houston
gxu@homer.phys.uh.edu